Homogeneous Catalysis by Metal Complexes

VOLUME II

Activation of Alkenes and Alkynes

Homogeneous Catalysis by Metal Complexes

VOLUME II
Activation of Alkenes and Alkynes

M. M. TAQUI KHAN
Department of Chemistry
Nizam College
Osmania University
Hyderabad, India

ARTHUR E. MARTELL
Department of Chemistry
Texas A & M University
College Station, Texas

ACADEMIC PRESS New York and London   1974
A Subsidiary of Harcourt Brace Jovanovich, Publishers

CHEMISTRY

ACADEMIC PRESS, INC.
111 Fifth Avenue, New York, New York 10003

United Kingdom Edition published by
ACADEMIC PRESS, INC. (LONDON) LTD.
24/28 Oval Road. London NW1

Library of Congress Cataloging in Publication Data

Taqui Khan, M   M
    Homogeneous catalysis by metal complexes.

    Includes bibliographical references.
    CONTENTS: v. 1.   Activation of small inorganic
molecules.–v. 2.   Activation of alkenes and alkynes.
    1.   Catalysis.   2.   Metal catalysts.   3.   Complex
compounds.   I.   Martell, Arthur Earl, Date
joint author.   II.   Title.
QD505.T36   546   72-9982
ISBN 0-12-406102-8 (v.2)

PRINTED IN THE UNITED STATES OF AMERICA

Contents

Preface

During the past twenty-five years the development of the field of coordination chemistry has gone through several general yet discernible phases. After a classic beginning based largely on descriptive and stereochemical studies, a more physical approach developed in which quantitative equilibrium studies led to the understanding of the thermodynamics of complex formation in solution. Subsequently the development of the ligand field theory and bonding concepts made possible successful correlations between electronic and magnetic spectra and the constitution and properties of coordination compounds. Presently, improvements in X-ray crystallographic techniques are providing a large body of structure–property correlations and producing a new level of understanding of coordination chemistry. Also, synthesis of new types of metal complexes of small molecules, many of which contain metal–carbon sigma and pi bonds, has opened new vistas in coordination chemistry, since these compounds are frequently catalysts or intermediates in the synthesis of new organic and organometallic compounds. In addition to their general applications to homogeneous catalysis, the complexes of nitrogen and oxygen in particular are of interest as models for biological oxidation and nitrogen fixation.

The purpose of this two-volume work is to review and systematize the chemistry of reaction of metal ions with small molecules, and to include chapters on hydrogen, oxygen, nitrogen, carbon monoxide, nitric oxide, and the alkenes and alkynes. Because many coordination chemists are primarily interested in the inorganic complexes, the decision was made to publish the metal ion activation of the small diatomic molecules as Volume I. Metal ion activation of alkenes and alkynes comprises Volume II. While the subject of metal ion activation of alkenes and alkynes is of vital importance to the field of coordination chemistry, it is broader in scope and would be of interest to organic and organometallic chemists as well as to coordination chemists.

As with many monographs, this work is the result of an earlier attempt to review and systematize the subject of interest for our own purposes. This effort began about seven years ago with a review of metal ion activation of molecular oxygen, a general subject closely related to previous research work

vii

on which we had collaborated. Completion of this review led over the next few years to the development of similar reviews on the activation of hydrogen, carbon monoxide, and unsaturated hydrocarbons. In the initial phase, the subject of nitrogen activation was not given sufficient weight to justify a full chapter, however the rapid growth of that field over the past few years led to the development of a nitrogen chapter that is now the most extensive in the work. Similar recent developments in nitrosation reactions have led to a chapter which is greatly expanded over its original concept; however in this case the field does not seem to have developed sufficiently to justify a treatment comparable to the other subjects.

Although the work began as a review for our own use only, we were prevailed upon by friends to prepare the work for publication. In a treatment of this nature, it is impossible to be expert in all phases of the field. Thus first-hand knowledge of metal ion activation of oxygen and nitrogen does not provide much insight into the fine points of reactions such as catalytic hydrogenation and nitrosation. In areas outside our range of expertise we have tried to present reactions and interpretations in a manner that seems reasonable to us. If our points of view do not coincide with those who are more experienced with the reactions under consideration, we hope for a charitable judgment of our treatment. We hope we have managed to present interpretations that are sufficiently original to make positive contributions to the subject at hand.

We have used our own preferred conventions for representation of covalent and coordinate bonds of organometallic compounds and metal complexes. We have related this to a consistent distribution of formal charges in the formulas used to represent these compounds. Details of the method employed have been presented in Appendix I. It is our hope that the readers will consider this approach to be both reasonable and satisfying. In any case one of its strong points is the designation of all formal charges for metal ions and ligand atoms, and where reasonably possible, their locations. While the formal charges assigned to the metal ions in many complexes seem to deviate widely from the accepted oxidation states, it is believed that the formalism employed corresponds more closely to (or reflects more satisfactorily variations in) the true charge distribution. In any case it is felt that our methods allow consistent and complete electron bookkeeping for the formulas under consideration.

We express our thanks to all those who provided valuable assistance in the preparation of the manuscript. Reviews of various chapters by the following professional friends and associates were particularly helpful: Professor Minoru Tsutsui (nitrogen), Professor Gordon Hamilton (oxygen), Dr. Elmer Wymore (hydrogen), and Dr. J. L. Herz (oxygen). Thanks and appreciation are also extended to Dr. R. Motekaitis and Dr. Badar Taqui Khan for assistance in proofreading the typescript and the galleys, to Robert M. Smith for literature searches, and to Mary Martell for editorial assistance and for typing the several versions of the manuscript that were produced over the long period of time during which this work developed.

M. M. Taqui Khan
Arthur E. Martell

Contents of Volume I

Introduction

Recent extensive investigations of the chemistry of σ- and π-bonded metal complexes of unsaturated organic compounds have not only opened up many new synthetic pathways, but have made possible detailed mechanistic studies through the isolation and study of many complex intermediates. Advances in ligand-field theory and molecular orbital concepts of π and σ bonding to metals have greatly aided in our understanding of the structures and re-activities of these intermediates, and have greatly aided in development of new concepts of reaction mechanisms and the development of new reaction types.

Since metal ions and complexes are catalysts for many reactions of alkenes and alkynes, the remarkable advances in the chemistry of their complexes have greatly accelerated the understanding of new mechanisms of homo-geneous catalysis and have promoted the development of many new catalytic processes for reactions of hydrocarbons. These improvements in theory and practice are expected to be applicable to the field of heterogeneous catalysis, since alkenes and alkynes are believed to be bonded to metal surfaces by the same types of σ and π bonds as are encountered in homogeneous systems.

This volume deals mainly with the activation of double (—C=C—) and triple (—C≡C—) bonds in a variety of compounds by metal ions and metal complexes. The examples selected include only homogeneous reactions in-volving metal ions and metal complexes in solution. The reactions discussed in detail include isomerization, hydrogenation, hydroformylation and hydrosil-ation, oxidation and hydration, Ziegler polymerization and oligomerization, multiple insertion reactions, and cyclic condensation on metallic centers.

Since most examples of π-bond activation involve the formation of labile

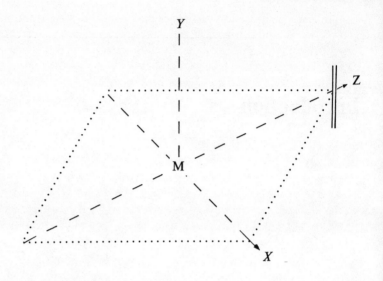

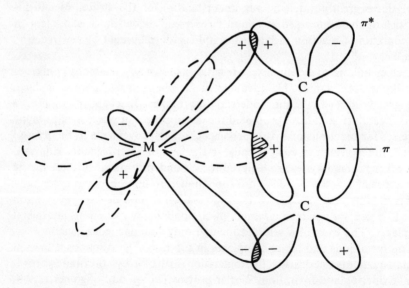

Fig. 1. Bonding in square-planar d^8 metal–olefin complexes.

π complexes, a discussion of the bonding in metal–alkene and metal–alkyne complexes is appropriate at this stage. According to the Dewar-Chatt-Duncanson theory [1,2] an alkene is bound to a d^8 metal ion with the plane of the alkene at right angles to the metal–alkene bond. The confirmation of perpendicular coordination of an alkene to a metal ion comes from the X-ray structure of Zeise's salt, $[Pt(Cl)_3C_2H_4]^-$, whereby it has been found [3,4] that the ethylene plane is at right angles to its coordinate bond and is symmetrically disposed about it (Fig. 1). Additional X-ray structures of a number of metal–alkene and metal–alkyne complexes have been described by Gusev and Struchkov [5]. In a bonding scheme of the type illustrated in Fig. 1, a metal–alkene σ bond is formed by the donation of a pair of electrons from the bonding π orbital of the alkene to a suitable hybrid orbital (e.g., dsp^2) of the d^8 metal ion. This is followed by π-bond formation through back donation of electrons from the filled pd orbitals of the metal ion to empty antibonding π^* orbitals on the alkene (Fig. 1). According to Denning et al. [6] this back donation may contribute more to the metal–alkene bond strength than does the σ bond.

A molecular orbital energy-level diagram incorporating the above concepts is presented in Fig. 2 [8]. If the coordinate system of Fig. 1 is taken (coordination of the olefin along the Z axis, with the olefin and three ligands coordinated to the metal ion in the XZ plane), the ordering of the d levels becomes $d_{z^2} > d_{xz} > d_{x^2-y^2} > d_{xy} > d_{yz}$ [8]. The σ-bonding orbitals on the ligands other than the olefin have been omitted to improve clarity. The formation of a dative π bond from the metal ion to the olefin can take place by back donation of electrons from d_{xz} or d_{yz} levels or a suitable combination of both. Such a description allows free rotation of the alkene around the metal–alkene σ bond, as evidenced by the NMR spectra of metal–alkene complexes [7]. The stability of a given d^8 metal–alkene complex depends on Δ, the separation between the highest filled and the lowest empty molecular orbitals [8]. This separation increases in the order nickel(II) < palladium(II) < platinum(II). For $PtCl_4^{2-}$ and $PdCl_4^{2-}$, Δ has been measured as 23,450 cm^{-1} and 19,150 cm^{-1}, respectively [9]. However, in metal–alkene complexes, this separation is difficult to determine purely on the basis of $d–d$ transitions because of complications due to charge-transfer contributions. The same general trend in Δ may be expected in this case as in the other square-planar complexes of palladium(II) and platinum(II). The Pt(II) complexes have the maximum stability of the metal ions with d^8 electron configuration, because of the greater spread of the $5d$ orbital energies as compared to those with $4d$ and $3d$ orbitals [palladium(II) and nickel(II)] and because of greater overlap of the $5d_{xz}$, $5d_{yz}$ hybrid orbitals with the π^* orbitals of the alkene. The stability constants for a variety of platinum(II) complexes have been given by Hartley [8], to whose review the reader is referred for further details. Nickel(II) and

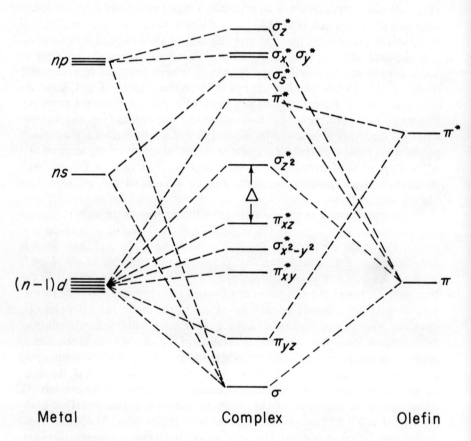

Fig. 2. Molecular orbital diagram for the metal–olefin complex of a d^8 square-planar metal ion.

palladium(II) alkene complexes are far less stable than those of platinum(II) and can act as catalytic species in a variety of reactions. Rhodium (I) complexes are more reactive than iridium(I) complexes for the same reason.

Metal–alkene complexes of ions having coordination number 6 are usually labile and most of them exist only as reaction intermediates. Transition metal ions having octahedral geometry are usually associated with higher oxidation states and in such cases the effective back donation from the t_{2g} orbitals is considerably reduced. This is especially the case with metal ions having either empty or partially empty t_{2g} levels such as titanium(III), vanadium(III), vanadium(II), and chromium(III), which do not have sufficient electrons for back-donation to the π^* levels of the alkene. In such cases metal–alkene complexes formed in solution are very labile and take part in subsequent reactions, most noteworthy of which are the polymerization reactions of the Ziegler-Natta catalysts (Chapter 6, Section VII).

Alkynes are coordinated to d^8 metal ion in much the same manner as are alkenes, with the molecule at right angles to the metal–alkyne bond. This has been confirmed with the X-ray structure of dichloro(p-toluidine)di-t-butyl-acetyleneplatinum(II) [10]. For alkyne complexes, two equivalent bonding concepts have been employed: one in which the alkyne may be assumed to occupy two coordination positions on the metal ion with the formation of a three-membered unsaturated ring, and the other in which a σ and a dative π bond are formed with the alkyne.

Nelson et al. [11,12] have carried out molecular orbital calculations on $(PPh_3)_3Pt(CH_3C{\equiv}CCH_3)$ and $(PPh_3)_2Pt(CH_3C{\equiv}CCH_3)$ in square-planar and trigonal (pseudotetrahedral) configurations, respectively, and came to the conclusion that the hybrid orbital for the square-planar complex is p^2d^2 and that of the trigonal hybrid p^2d. The bonding scheme for the pseudotetrahedral complex $(PPh_3)_2Pt(CH_3C{\equiv}CCH_3)$ is presented in Fig. 3. The alkyne is coordinated to the metal ion with the C—C bond at right angles to the metal–alkyne bond. In such an arrangement two donor bonds are formed from alkyne to platinum(II) by the overlap of the π_{xy} and $\pi_{z\parallel}$ (perpendicular to π_{xy}) orbitals of the alkyne with the p^2d and the $(d_{xz} + d_{yz})$ hybrid on platinum(II), respectively. This overlap may be seen in an edge-view of the complex (Fig. 3), where the cylindrical π cloud about the C—C σ bond in the alkyne forms donor bonds with platinum(II). The formation of the donor bonds from alkyne to platinum(II) is complemented by back donation of electrons from the filled $d_{x^2-y^2}$ and the $(d_{xz}-d_{yz})$ pair, respectively, on the metal ion to empty π_{xy}^* and $\pi_{z\parallel}^*$ empty orbitals, respectively, of the alkyne. One such dative π bond is depicted in the top view of the complex in Fig. 3. The calculated electron density on acetylenic carbons due to back donation was found to be much higher than that of ethylene. This led [13] to the conclusion that acetylene is a better π acceptor than ethylene. Practically the same conclusions

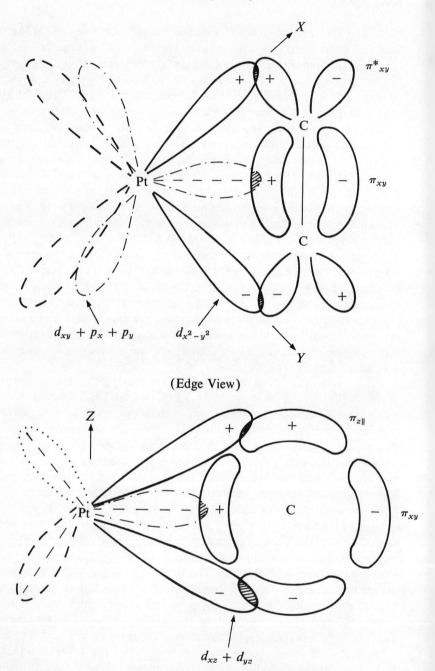

(Top View)

π^*_{xy}

π_{xy}

$d_{xy} + p_x + p_y$

$d_{x^2-y^2}$

(Edge View)

$\pi_{z\parallel}$

π_{xy}

$d_{xz} + d_{yz}$

Fig. 3. Bonding in platinum–alkyne complexes.

6

about the nature of the metal–alkyne bond is reached in the square-planar configuration [11]. The stabilities of Ni(II)–alkyne and Pd(II)–alkyne complexes are expected to be much lower than those of Pt(II) complexes, which suggests a reason for the reactivity of these ions as catalysts in the insertion reactions of alkynes, especially the oligomerization and polymerization of alkynes (Chapter 6, Section VI).

Migration of π Bonds

2

The main types of unsaturated molecules considered in this chapter are mono-olefins, conjugated dienes, and nonconjugated dienes separated by one or more methylene groups. Metal-catalyzed double-bond migration in mono-olefins and nonconjugated dienes giving rise to position isomers most probably involves activation of the π bond through the formation of π complexes of the metal ions. Migration of the π bond to suitable internal positions is sometimes accompanied by geometric inversion. The ratio of *trans* to *cis* isomers obtained in such cases is usually governed by their thermodynamic equilibrium, with formation of the more stable *trans* isomer predominating. Metal carbonyls (usually those of cobalt and iron), metal ions of the platinum group, and monovalent silver compounds are effective catalysts in π-bond isomerization, because of their ability to form complexes with the olefinic bond. The relative reactivities of these metals probably depend on the kinetic and thermodynamic stabilities of the π complexes formed.

In this chapter, catalytic isomerization of olefins by metal carbonyls is treated first, because of the widespread importance of such reactions in hydroformylation of olefins and because of the relationship of isomerization to other reaction types. The mechanisms of these reactions are helpful to a certain extent in understanding the catalytic role of platinum group metals in π-bond migration. For more detailed information the reader is referred to a review by Orchin [14] of the catalytic isomerization of olefins by transition metal complexes.

I. π-Bond Migration Catalyzed by Metal Carbonyls

A. COBALT CARBONYLS

The catalytic activity of dicobalt octacarbonyl in the isomerization of double bonds under hydroformylation conditions in the presence of carbon monoxide and hydrogen has been described by Wender et al. [15,16]. When 1-hexene was heated at 110°C for 1 hour in the presence of dicobalt octa-carbonyl, under a carbon monoxide pressure of 133 atm, 9% of the 1-hexene was converted to the internal olefins, 2-hexene and 3-hexene, the rate of isomerization increasing with temperature and time of heating. When the same reaction was conducted at 150°C for 5 hours, a 61% combined yield of 2-hexene and 3-hexene was obtained. It has been shown subsequently by Heck and Breslow [17] that cobalt hydrocarbonyl is produced from dicobalt octacarbonyl in the reaction mixture and is the species that takes part in hydroformylation and isomerization reactions. π-Allyl complexes are readily obtained by the reaction of cobalt hydrocarbonyl with butadiene or substituted butadienes by transfer of a hydride ion to one of the methylene groups, as indicated by Eq. (1). The methyl π-allyl complex thus obtained is a mixture

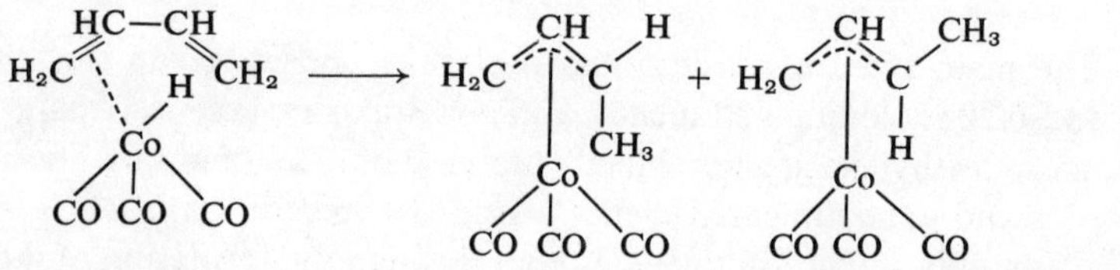

$$\text{(1)}$$

of two isomers, the *anti* form, **2a**, and the *syn* form, **2b**, with the equilibrium favoring the latter. Thus hydride addition to the coordinated diene in **1** to give a coordinated π-allyl complex gives rise to an unequal distribution of the two possible isomers. Geometrical isomers of this type are involved in most of the metal-catalyzed reactions of internal olefins and conjugated diolefins and will be discussed below under a separate heading.

Extensive work on the isomerization of olefins by cobalt hydrocarbonyl has been described by Karapinka and Orchin [18], Heck and Breslow [17], and by Johnson [19]. Karapinka and Orchin [18] reported catalysis of the isomerization of 1-pentene by cobalt hydrocarbonyl under a nitrogen atmosphere at 25°C with a 25:1 ratio of substrate to catalyst. Under these conditions, 52% of 1-pentene was isomerized to a mixture of *trans*- and *cis*-2-pentene. Investigation of the reaction in a carbon monoxide atmosphere showed that increasing the carbon monoxide concentration caused a decrease in the rate of isomerization. Markownikoff addition of cobalt hydrocarbonyl to the olefin, resulting in the formation of an alkylcobalt tetracarbonyl complex has been suggested [18]. The latter then breaks down to form the isomerized olefin and regenerate the cobalt hydrocarbonyl, as indicated by reactions (2) and (3). Similar reactions were suggested by Sternberg and Wender [20] for the isomerization of α-olefins by cobalt hydrocarbonyl.

$$CH_3(CH_2)_2CH=CH_2 + HCo(CO)_4 \longrightarrow CH_3(CH_2)_2\overset{\overset{\displaystyle CH_3}{|}}{C}HCo(CO)_4 \qquad (2)$$

$$CH_3(CH_2)_2\overset{\overset{\displaystyle CH_3}{|}}{C}HCo(CO)_4 \longrightarrow CH_3CH_2CH=CHCH_3 + HCo(CO)_4 \qquad (3)$$

The inhibition of the rate of isomerization by increased carbon monoxide pressure, and the nature of the intermediates involved in alkyl–π-complex equilibria giving rise to isomerization, was thoroughly investigated by Heck and Breslow [17]. The same end products have been reported [17] for a hydroformylation reaction (addition of —CHO and H across a double bond) starting with 1-pentene or 2-pentene, the only difference between them being a slower rate of reaction with 2-pentene as compared to 1-pentene. The active catalytic intermediate participating in the isomerization reaction was postulated [17] to be the coordinately unsaturated species $HCo(CO)_3$ obtained by a reversible dissociation of $HCo(CO)_4$ according to reaction (4). The observed

$$HCo(CO)_4 \rightleftharpoons HCo(CO)_3 + CO \qquad (4)$$

inverse dependence of the rate of isomerization on carbon monoxide concentration [13,14] is in accord with this dissociation reaction. The species $HCo(CO)_3$ then forms a π-olefin complex with 1-pentene or 2-pentene, as shown in the reaction scheme of 3–7. The mono-olefin complexes 3 and 6 are

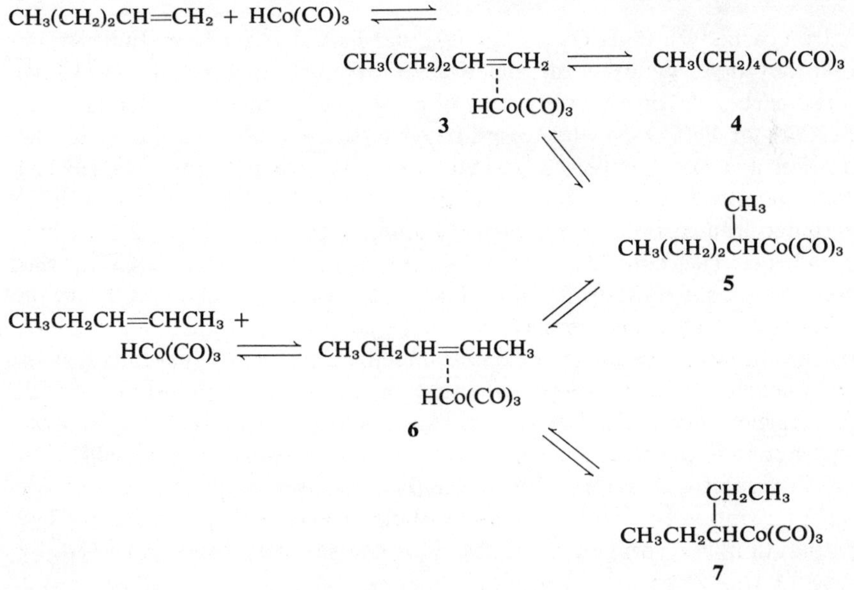

$$CH_3(CH_2)_2CH=CH_2 + HCo(CO)_3 \rightleftharpoons$$

$$CH_3(CH_2)_2CH\overset{\vdots}{=}CH_2 \rightleftharpoons CH_3(CH_2)_4Co(CO)_3$$
$$HCo(CO)_3$$

3   **4**

$$CH_3(CH_2)_2\overset{\overset{\displaystyle CH_3}{|}}{C}HCo(CO)_3$$

5

$$CH_3CH_2CH=CHCH_3 + HCo(CO)_3 \rightleftharpoons CH_3CH_2CH\overset{\vdots}{=}CHCH_3$$
$$HCo(CO)_3$$

6

$$CH_3CH_2\overset{\overset{\displaystyle CH_2CH_3}{|}}{C}HCo(CO)_3$$

7

in equilibrium with alkylcobalt tricarbonyl complexes **4**, **5**, and **7**, respectively. Thus, starting with 1-pentene, 2-pentene can be obtained through intermediates **3**, **5**, and **6**. A similar reversible reaction involving metal π-complex–metal alkyl formation and dissociation is the dissociation of ethylcobalt tetracarbonyl into ethylene [17]. Injection of a solution of ethylcobalt tetracarbonyl into the heated inlet of a gas chromatographic apparatus at 220°C gave an 80% yield of ethylene, as shown in Eq. (5). Johnson [19] postulated a

$$C_2H_5Co(CO)_4 \rightleftharpoons C_2H_4 + HCo(CO)_4 \tag{5}$$

similar scheme for the stepwise migration of double bonds in the isomerization of 1-pentene under hydroformylation conditions. The postulated [19] stepwise migration of the double bond under oxo conditions may be due to the tendency of the olefin to form a straight-chain aldehyde. Further details of the oxo reaction are considered in Chapter 3, Section II.

Taylor and Orchin [21] have provided further evidence for the 1,2-addition–elimination mechanism by experiments on the isomerization of propene-d_6 with hydridocobalt tetracarbonyl, $HCo(CO)_4$, as catalyst. In order to avoid solvent effects and to more closely control the concentration of carbon monoxide, the isomerization was conducted in the gaseous phase. An appreciable incorporation of hydrogen at the 2-carbon atom of propene was observed [21], indicating isomerization by 1,2-addition–elimination of $HCo(CO)_4$.

B. IRON CARBONYLS

Iron carbonyls, $Fe_3(CO)_{12}$, $Fe_2(CO)_9$, and $Fe(CO)_5$ have been studied more than any other catalysts in isomerization of mono-olefins and polyolefins. These carbonyls form a wide variety of π complexes with olefins having either $Fe(CO)_4$ or $Fe(CO)_3$ groups coordinated to the π bonds, depending on the number of π bonds involved. In each case, one carbonyl group is displaced from the metal carbonyl by each olefinic π bond formed. Rehlen *et al.* [22] prepared iron carbonyl complexes of butadiene, isoprene, and 2,3-dimethylbutadiene. The complex obtained with butadiene, $(C_4H_6)Fe(CO)_3$, was assigned a configuration having σ iron–carbon bonds. Hallam and Pauson [23] reinvestigated this complex and, on the basis of reduction and ozonolysis reactions, and of magnetic and spectral studies, postulated that butadiene is an intact moiety in the complex and that the iron atom is equidistant from all four carbon atoms. This conclusion was confirmed later by X-ray studies. The π bond in the butadiene–iron tricarbonyl complex was postulated [23] to be of the same type as that of ferrocene. This discovery provided incentive for the preparation and study of many diolefin and π-allyl complexes. The preparation and properties of these complexes have been reviewed by

Fischer and Werner [24], Green and Nagy [25], and by Guy and Shaw [26], to whose publications the reader is referred for further details.

The rearrangement of iron carbonyl π complexes to isomerized products depends on the nature of olefin, solvent, and temperature. The choice of the appropriate iron carbonyl catalyst, $Fe_3(CO)_{12}$, $Fe_2(CO)_9$, or $Fe(CO)_5$, depends on the thermal stability of the iron carbonyl under investigation. The carbonyls $Fe_3(CO)_{12}$ and $Fe_2(CO)_9$ are used in the temperature range $60°-80°C$, while $Fe(CO)_5$ is effective above $100°C$. The ratio of the various isomers obtained by rearrangement of π bonds is controlled by their relative thermodynamic stability. In cases of geometric isomers, the more stable *trans* isomer is always formed in higher concentrations.

Manuel [27] studied the isomerization of a series of monoolefins with iron pentacarbonyl and reported that the ratio of the isomers obtained conforms to the normal distribution expected on the basis of the relative thermodynamic stabilities. Internal olefins react more slowly with iron pentacarbonyl than do the terminal (α-) olefins, and trisubstituted olefins are even less reactive than the internal olefins. The order of reactivity of the various olefins is terminal olefin > *cis*-internal olefin > *trans*-internal olefin > trisubstituted olefin. The rate of isomerization is inhibited by increasing carbon monoxide pressure. The addition of small quantities of polar substances, such as acetone, 1,2-dimethoxyethane, and ethanol, increase the rate of isomerization. Addition of acetic acid, however, retards the rate. Iron pentacarbonyl is unreactive for the isomerization of 1-hexene below $100°C$, but is an effective catalyst in the presence of pyridine. This base (in the presence of alcohols) seems to help the disproportionation of iron carbonyls with the formation of hydrido–metal carbonyl complexes [28], which probably act as the reactive catalytic intermediates. This conclusion is supported by the characteristic red color of these hydrides observed [29] in the reaction mixture. Catalysis of the isomerization of 1-hexene by the iron carbonyl hydride anion, $[H_2Fe_2(CO)_8]^{2-}$, reported by Sternberg *et al.* [30,31] provides further support for the postulation of hydrido intermediates in these reactions.

Two mechanisms have been proposed [27,29] for the isomerization of mono-olefins by iron pentacarbonyl. According to the first mechanism, a metal carbonyl hydrido-π-complex is considered in equilibrium with an alkyl–metal carbonyl complex. This mechanism is analogous to that proposed by Heck and Breslow [17] for the isomerization of mono-olefins by hydrido-cobalt tetracarbonyl, $HCo(CO)_4$. The source of the hydride donor in the iron carbonyl hydride π complex, compound **8**, is considered to be derived from a compound other than the olefin undergoing reaction. This source of hydrogen may very well be one of the bases such as acetone, pyridine, or 1,2-dimethoxyethane, that are generally effective in promoting isomerization. Also, the solvent may produce hydride by a redox mechanism, such as

Eq. (6). Geometric isomers may be formed by rotation of the CH_2R group about the single bond in the alkyl complex **9** as in Eq. (7).

$$CH_3CH_2OH + ML \longrightarrow CH_3CHO + HL + HM \qquad (6)$$

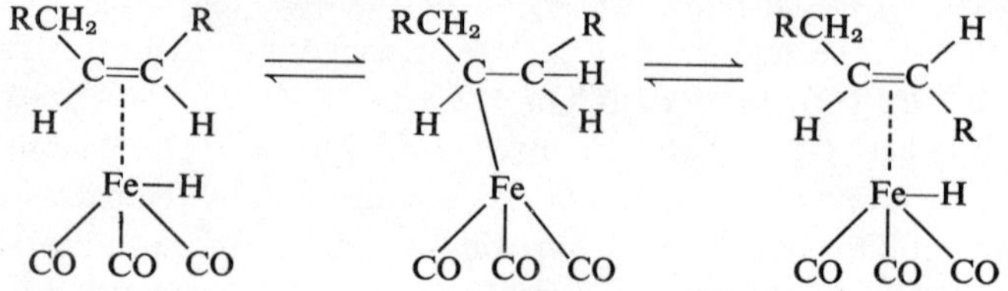

In an alternate mechanism outlined in Eq. (8), $Fe(CO)_5$ forms an olefin tetracarbonyl π complex, **11**. This is followed by abstraction of an allylic hydrogen to form a π-allyl iron tricarbonyl hydride (**12**). In this π-allyl complex, the original double bond loses its identity and the return of the hydride ion to a different carbon atom results in the isomerization of the olefin to give the π complex, **13**:

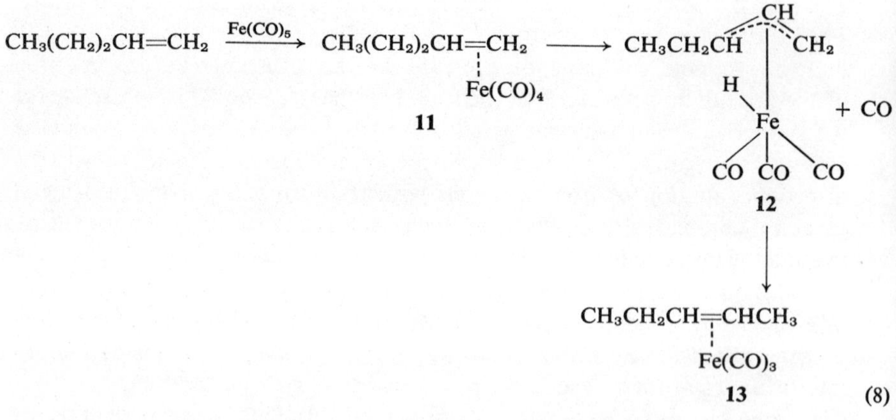

Support for the π-allyl mechanism comes from the ability of iron carbonyls to form π-allyl complexes from mono-olefins by abstraction of either a hydride ion or a halide ion. In the latter case a π-allyliron carbonyl halide is formed. Murdoch and Weiss [32] prepared π-allyliron carbonyl halides by the interaction of allyl halides and $Fe_2(CO)_9$. The halides 3-chloro-1-butene and 1-chloro-2-butene gave the same product, methyl-π-allylchloroiron tricarbonyl (**14**), as indicated by reaction sequence (9) [33]. Migration of the double bond through the formation of π-allyl intermediates [34] for the isomerization of allyl alcohol to propionaldehyde is indicated in Eqs. (10) and (11). According to the mechanism proposed, allyl alcohol forms a π complex, **15**, by reaction

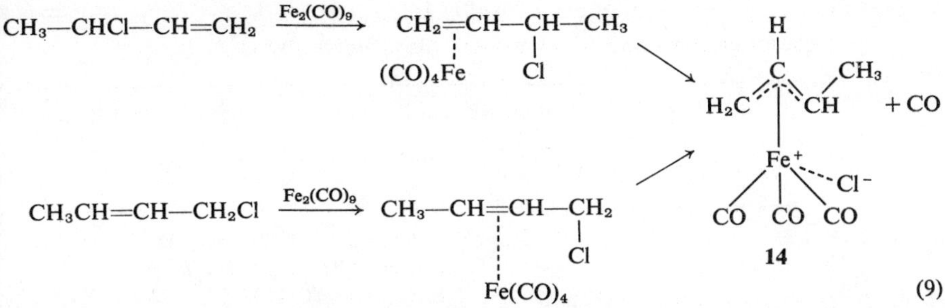

$$CH_3—CHCl—CH=CH_2 \xrightarrow{Fe_2(CO)_9} CH_2=CH—CH—CH_3$$

at centre: $(CO)_4Fe$, Cl

Structure at right:

$$
\begin{array}{c}
H \\
C \overset{CH_3}{\diagdown} \\
H_2C \diagup \overset{|}{\diagdown} CH \quad + CO \\
Fe^+ \\
\diagup \overset{|}{\cdots} Cl^- \\
CO \quad CO \quad CO \\
14
\end{array}
$$

$$CH_3CH=CH—CH_2Cl \xrightarrow{Fe_2(CO)_9} CH_3—CH=CH—CH_2$$

$$Cl$$

$$Fe(CO)_4 \tag{9}$$

with iron pentacarbonyl. This complex is converted to a π-allylhydridoiron tricarbonyl complex, **16**, by hydride ion abstraction from the —CH_2OH group. Isomerization of the hydroxy compound takes place by the return of the hydride ion to a different position in **17**. Ligand exchange then takes place with unreacted allyl alcohol and the displaced ligand fragment $CH_3—CH=CHOH$ tautomerizes to give propionaldehyde [Eq. (11)].

$$CH_2=CH—CH_2OH + Fe(CO)_5 \longrightarrow$$

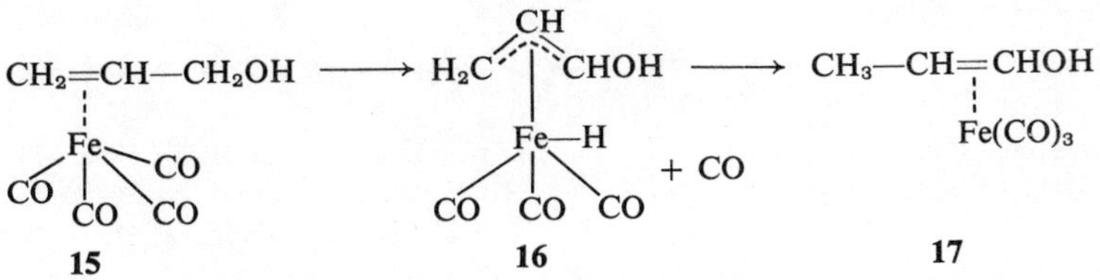

$$CH_2=CH—CH_2OH \longrightarrow H_2C \overset{CH}{\diagdown} CHOH \longrightarrow CH_3—CH=CHOH$$

beneath first: Fe with CO, CO, CO — **15**

beneath middle: Fe—H with CO, CO, CO + CO — **16**

beneath right: $Fe(CO)_3$ — **17**

$$(10)$$

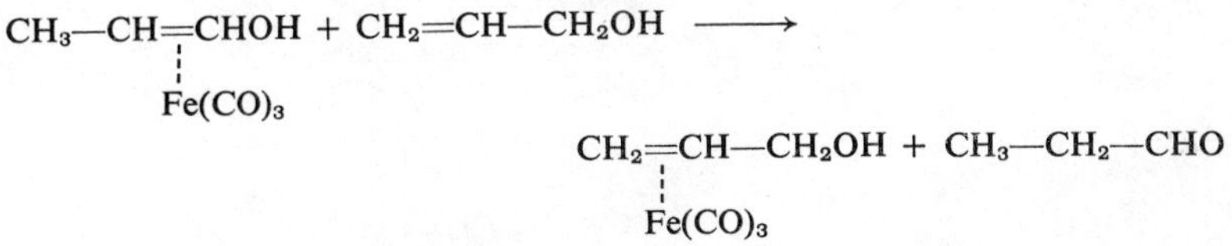

$$CH_3—CH=CHOH + CH_2=CH—CH_2OH \longrightarrow$$

$$Fe(CO)_3$$

$$CH_2=CH—CH_2OH + CH_3—CH_2—CHO \tag{11}$$

$$Fe(CO)_3$$

In the π-allyl mechanism for the isomerization of mono-olefins, rearrangement of the π bonds take place within the π-allyl complex. The migration of the π bond does not necessarily involve the formation of an intermediate olefin. It may be seen from Table I that 1-hexene is converted rapidly, although partially, to 3-hexene, even though 2-hexene reacts much more slowly. It is also seen that 2-methyl-1- and 2-pentenes are formed from 4-methyl-1-pentene despite the fact that the alkene of intermediate structure, 4-methyl-2-pentene, is relatively inert. The data thus seem to support the π-allyl mechanism for the isomerization of mono-olefins. Nevertheless, it is difficult to completely rule out the π-complex alkyl mechanism in favor of the π-allyl mechanism. One of the main differences between the two mechanisms lies in

TABLE I

ISOMERIZATION OF END AND INTERNAL MONO-OLEFINS WITH $Fe_3(CO)_{12}$[a]

No.	Substrate	Time of the reaction (hours)	Products	Composition %
1	1-Hexene $CH_2{=}CH(CH_2)_3CH_3$	4	$CH_3CH{=}CH(CH_2)_2CH_3$ (*cis*) $CH_3CH{=}CH(CH_2)_2CH_3$ (*trans*) $CH_3CH_2CH{=}CHCH_2CH_3$	16.2[b] 57.6 25.0
2	4-Methyl-1-pentene $CH_2{=}CHCH_2\overset{\displaystyle CH_3}{\underset{\displaystyle |}{C}}H{-}CH_3$	19	$CH_2{=}\overset{\displaystyle CH_3}{\underset{\displaystyle |}{C}}{-}(CH_2)_2CH_3$ $CH_3{-}\overset{\displaystyle CH_3}{\underset{\displaystyle |}{C}}HCH{=}CHCH_3$ $CH_3CH_2CH{=}C(CH_3)_2$	13.0 13.0 74.0
3	Vinylcyclohexane 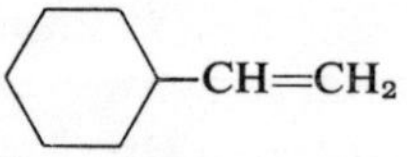 ⬡$-CH{=}CH_2$	19	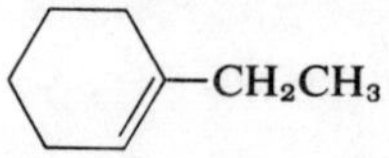 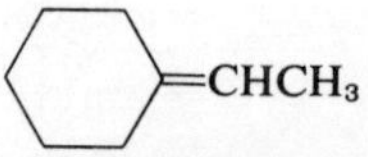 ⬡$-CH_2CH_3$ ⬡${=}CHCH_3$	—
4	2-Hexene (82% *cis*) $CH_3CH{=}CH(CH_2)_2CH_3$	19	$CH_3CH{=}CH(CH_2)_2CH_3$ (*cis*) $CH_3CH{=}CH(CH_2)_2CH_3$ (*trans*) $+ CH_3CH_2CH{=}CHCH_2CH_3$	74.0 25.0
5	*trans*-4-Methyl-2-pentene $CH_3\overset{\displaystyle H}{\underset{\displaystyle |}{C}}{=}\overset{\displaystyle CH_3}{\underset{\displaystyle |}{\underset{\displaystyle H}{C}}}{-}CHCH_3$	16	$CH_3CH{=}CH\overset{\displaystyle CH_3}{\underset{\displaystyle |}{C}}H{-}CH_3$[c] (starting material) $CH_3CH_2CH_2\overset{\displaystyle }{\underset{\displaystyle CH_3}{\underset{\displaystyle |}{C}}}{=}CH_2$ $CH_3CH_2CH{=}C(CH_3)_2$	95 2 2
6	*cis*-4-Methyl-2-pentene $CH_3\overset{\displaystyle CH_3}{\underset{\displaystyle }{C}}{=}\overset{\displaystyle }{\underset{\displaystyle H}{\underset{\displaystyle |}{C}}}{-}\overset{\displaystyle }{\underset{\displaystyle H}{\underset{\displaystyle |}{C}}}H{-}CH_3$	16	$CH_3CH{=}CH\overset{\displaystyle CH_3}{\underset{\displaystyle |}{C}}H{-}CH_3$[c] $CH_3CH_2CH_2\overset{\displaystyle }{\underset{\displaystyle CH_3}{\underset{\displaystyle |}{C}}}{=}CH_2$ $CH_3CH_2CH{=}C(CH_3)_2$	12 12 77
7	2-Methyl-2-pentene $CH_3CH_2CH{=}C(CH_3)_2$	—	No reaction	—

[a] From reference [27].
[b] The remaining products consist of less volatile unidentified substances.
[c] Equilibrium mixture of *cis* and *trans* forms.

the source of the hydride ion to form the hydridoiron carbonyl anion. In the π-complex–metal alkyl mechanism the hydride ion appears to come from the bases used as co-catalysts, whereas in the π-allyl mechanism the hydrogen is abstracted from the allylic proton present in the original olefin.

Roos and Orchin [35] reported the isomerization of allylbenzene (18) with $HCo(CO)_4$ and $DCo(CO)_4$ as catalysts. The catalytic isomerization of allyl-benzene proceeds rapidly and gives trans-propenylbenzene (19) as a major

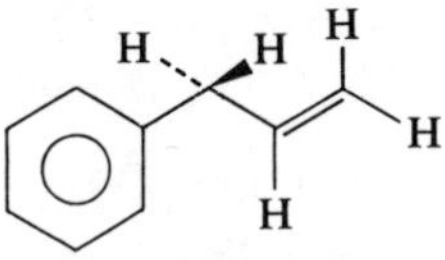

18 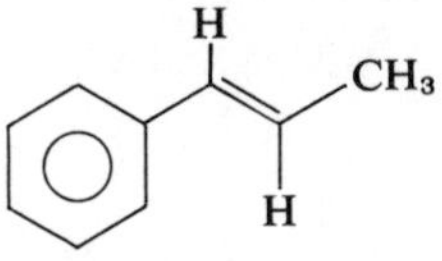   19

product. The $HCo(CO)_4$- and $DCo(CO)_4$-catalyzed reactions proceeded essentially with equal rates and there was very little (5.3%) incorporation of deuterium in the propenyl product. The results are thus neither in accord with the π-allyl mechanism nor the reversible hydride addition–elimination mechanism because both mechanisms predict a large isotope effect as a result of Co—H and Co—D bond cleavage and incorporation of deuterium in the isomerized product. An internal 1,3 (intramolecular) hydrogen shift has been proposed for the isomerization of allylbenzene but the details of such a mechanism await further investigations.

C. ISOMERIZATION OF DIOLEFINS

In an attempt to prepare a diene–iron tricarbonyl complex from 1,5-cyclooctadiene 20, King et al. [36], Manuel and Stone [37], and Arnet and Pettit [38], reported that the expected 1,5-cyclooctadiene–iron tricarbonyl complex was not obtained, and the product quantitatively formed was the complex of the conjugated 1,3-cyclooctadiene 21 as indicated in Eq. (12). The same type of isomerization was observed [38] with the following catalysts: bicycloheptadiene–iron tricarbonyl, cyclooctatetraene–iron tricarbonyl, and hexadiene–iron tricarbonyl complexes. These reactions represent the first examples of the isomerization of nonconjugated to conjugated dienes with metal carbonyls as catalysts. The unconjugated diene 1,4-pentadiene (22) reacts in a similar manner with $Fe_3(CO)_{12}$ [31] or $Fe(CO)_5$ [33,39,40] to

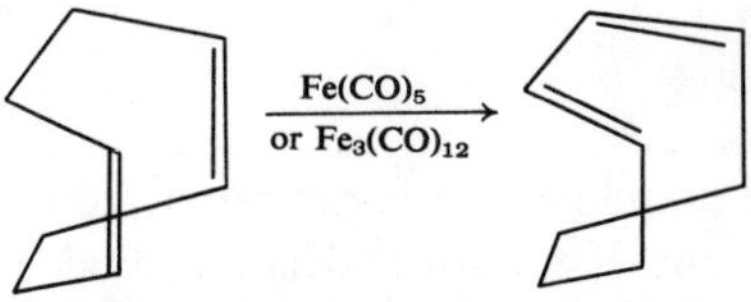

20   21

produce *trans*-1,3-pentadiene–iron tricarbonyl, **23**. Both *cis*- and *trans*-1, 3-pentadiene react with $Fe(CO)_5$ to give only the *trans*-1,3-pentadiene–iron tricarbonyl complex, **23**. Up to the present time no direct method is known for the preparation of the *cis*-1,3-pentadiene–iron tricarbonyl complex. The

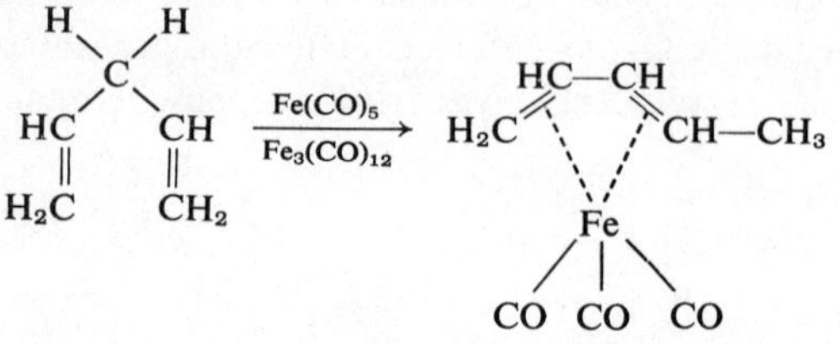

$$
\begin{array}{ccc}
\underset{\textbf{22}}{\text{22}} & \xrightarrow[\text{Fe}_3(\text{CO})_{12}]{\text{Fe(CO)}_5} & \underset{\textbf{23}}{\text{23}}
\end{array} \tag{13}
$$

unconjugated cyclic diene (**24**) 1,4-cyclohexadiene gave 1,3-cyclohexadiene–iron tricarbonyl (**25**) on treatment with iron pentacarbonyl [33,40]. 1,4-Dihydromesitylene (**26**) was converted to the 1,3-derivative **27** on treatment

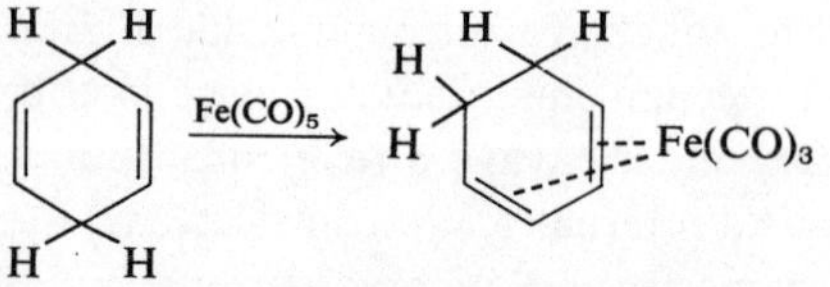

$$
\begin{array}{ccc}
\underset{\textbf{24}}{\text{24}} & \xrightarrow{\text{Fe(CO)}_5} & \underset{\textbf{25}}{\text{25}}
\end{array} \tag{14}
$$

with $Fe_3(CO)_{12}$ [36]. Similarly, *dl*-limonene, **28**, gave [36] terpinene **29** on

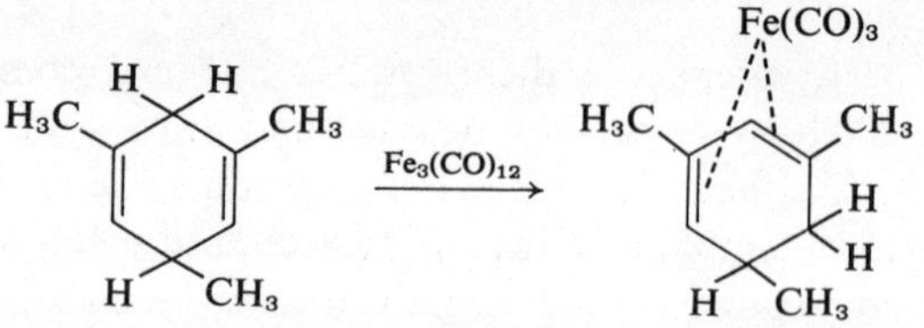

$$
\begin{array}{ccc}
\underset{\textbf{26}}{\text{26}} & \xrightarrow{\text{Fe}_3(\text{CO})_{12}} & \underset{\textbf{27}}{\text{27}}
\end{array} \tag{15}
$$

isomerization with $Fe_3(CO)_{12}$ as a catalyst.

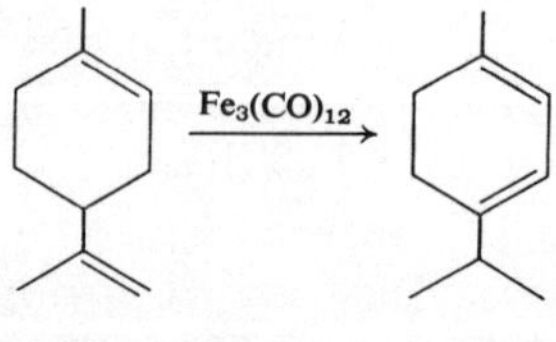

$$
\begin{array}{ccc}
\underset{\textbf{28}}{\text{28}} & \xrightarrow{\text{Fe}_3(\text{CO})_{12}} & \underset{\textbf{29}}{\text{29}}
\end{array} \tag{16}
$$

Rearrangement of double bonds is not restricted to nonconjugated dienes. Cases of rearrangement of substituted conjugated dienes with iron pentacarbonyl as catalyst were recently reported by Pettit and Emerson [39].

Thus 2,5-dimethyl-2,4-hexadiene (30) reacts with iron pentacarbonyl to give *trans*-2,5-dimethyl-1,3-hexadiene–iron tricarbonyl (31). The same product is also obtained on the treatment of 2,5-dimethyl-1,5-hexadiene (32) with iron pentacarbonyl. The conjugated diene, 4-methyl-1,3-pentadiene (33) re-arranged to *trans*-2-methyl-1,3-pentadiene–iron tricarbonyl (34). Reactions

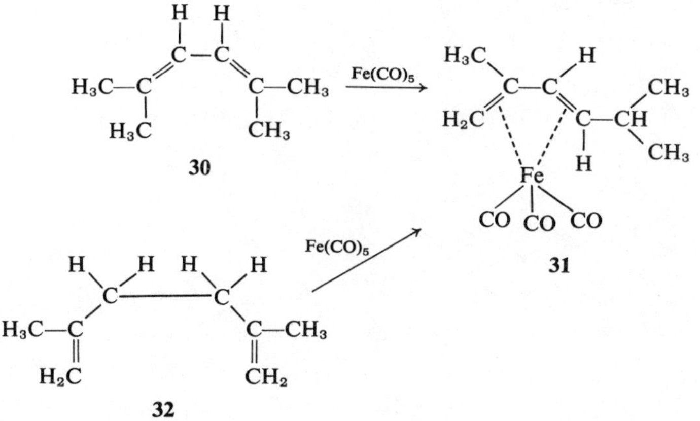

(17)

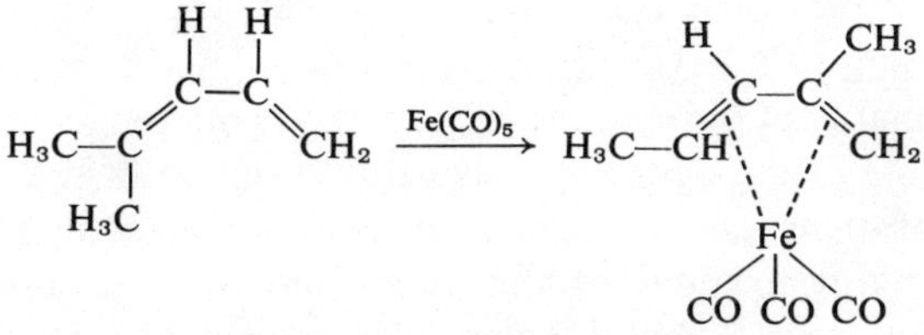

(18)

(17) and (18) indicate that the direction of isomerization in the case of conju-gated dienes may be controlled by steric effects, which influence the thermo-dynamic stabilities of the isomerized products. The more stable reaction products 31 and 34 have their substituent groups in the *trans* position. This is probably due to the steric requirement imposed by the iron tricarbonyl group on the two double bonds, which are held in an approximately coplanar posi-tion and in a cissoid conformation. Thus the rearrangement appears to be influenced mainly by the new steric requirements of the substituents imposed by the planar diene structure and the coordinated metal carbonyl moiety.

Pettit *et al.* [33] proposed a mechanism for the $Fe(CO)_5$-catalyzed isomeri-zation of nonconjugated dienes whereby an iron tricarbonyl π complex (e.g., 35), involving one or two π bonds, is first formed. The iron then abstracts an allylic hydrogen to form the intermediate π-allylhydridoiron tricarbonyl complex 36, which rearranges to a conjugated diene–iron tricarbonyl com-plex, 37. The driving force for the rearrangement appears to be the greater thermodynamic stability of the metal π chelate of the conjugated rearranged isomer, **37**, as compared to the π complex of the 1,4-diene, **35**. Another factor

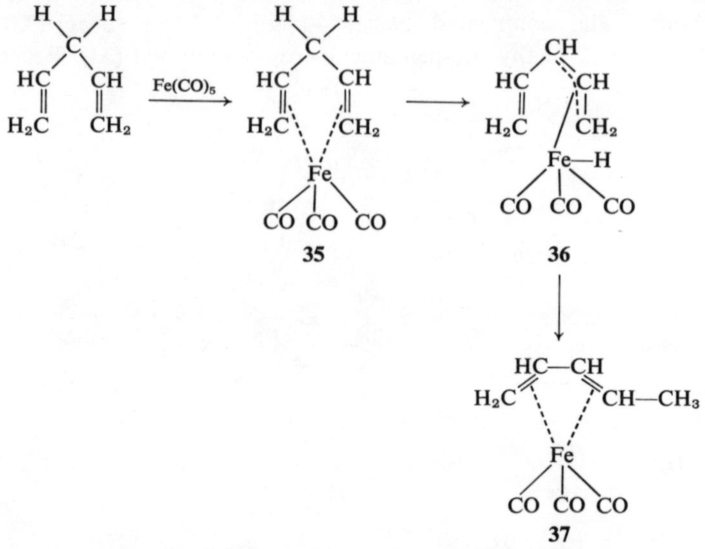

in the driving force for the rearrangement of double bonds appears to be the configurational fit of the π bonds of the olefin about the metal atom. Thus the "chelate" ring of the conjugated diene **37** seems to fit the coordination requirement of the metal ion better than the ring formed by the unconjugated diene. A striking example of this effect [43] is the isomerization of 1,3-cyclooctadiene to 1,5-cyclooctadiene with rhodium(I) as a catalyst. Rhodium(I) forms a very stable π-diolefin complex with 1,5-cyclooctadiene, which appears to be favored over the formation of the analogous π complex of the conjugated 1,3-cyclooctadiene. Although 1,3-cyclooctadiene is itself thermodynamically more stable than the 1,5-isomer, the rearrangement goes in the direction of the thermodynamically more stable complex. It is noted that the reverse isomerization reaction is catalyzed by iron carbonyls, as is indicated above [Eq. (12)].

Ogata and Misono [42] proposed a π-allyl mechanism for the isomerization of nonconjugated dienes, which is similar in most respects to that proposed by Pettit *et al.* [33]. The tricarbonyl moiety with the nonconjugated diene is assumed to involve only one double bond, to form a complex, **38**. Complex **38** then isomerizes to π-allylhydroiron tricarbonyl complex **39**. The rearrangement of **38** to **40** is similar to that of **35** to **37** in Eq. (19).

In nonconjugated dienes separated by several single bonds, an initial chelate structure as proposed by Pettit *et al.* [33] **35** may not be sterically probable. In such cases interaction of the catalyst with only one of the double bonds seems more feasible.

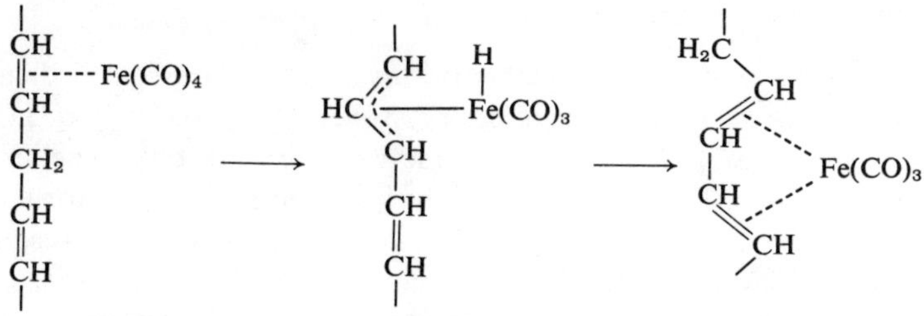

$$\text{(20)}$$

In the mechanisms proposed by Pettit *et al.* [33] and Ogata and Misono [42], the presence of allylic hydrogens seems to be essential for double-bond migration. Indirect evidence for this requirement comes from the hydrolysis of allyliron tricarbonyl cations [34], formed from the corresponding diene–iron tricarbonyl and a strong protonic acid. The hydrolysis of the π-1,1-dimethylallyliron tricarbonyl cation (41) results in addition of hydroxide ion to give 3-methyl-1-butene-3-ol, 43. The 1-methylallyl cation 44, however, gives methyl ethyl ketone, 47, on hydrolysis. In the case of the π-1,1-dimethylallyliron tricarbonyl cation 41, there is no allylic hydrogen, and π-bond migration is not possible. In compound 44, however, the presence of allylic hydrogen permits the abstraction of the hydride ion by the $Fe(CO)_3$ moiety, with the subsequent isomerization of the double bond to give intermediates 45 and 46, which subsequently tautomerize to the product, methyl ethyl ketone, 47.

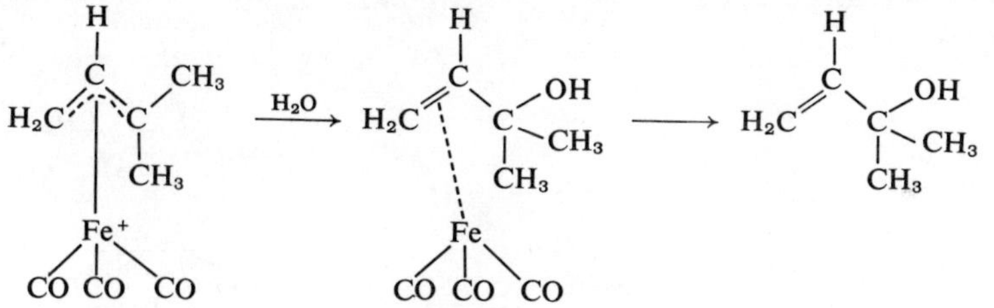

$$\text{(21)}$$

$$\text{(22)}$$

$$CH_3CH_2COCH_3$$

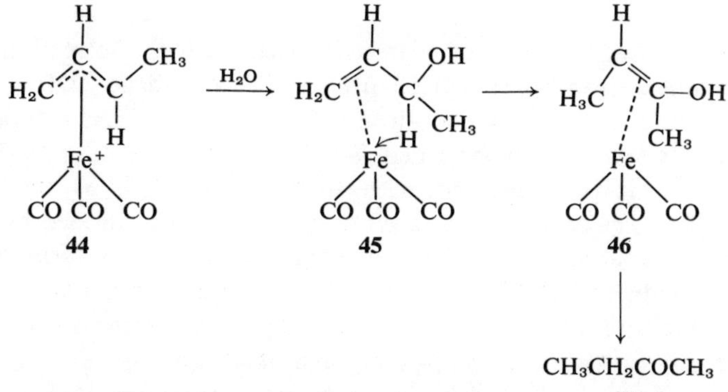

47

D. GEOMETRIC INVERSION

Several interesting cases of geometric inversion of conjugated dienes involving catalysis by $Fe(CO)_5$ have been reported by Emerson et al. [41] and Mahler et al. [44,45]. According to these investigators, butadiene–iron tricarbonyl (48) adds hydrogen chloride by cis addition [41] to form syn-1-methylallylchloroiron tricarbonyl, 51. The compound 51 was first assigned an anti configuration by Impastato and Ihrman [46]. On the basis of spectral and NMR evidence Emerson et al. [41] reported that the compound has the syn configuration. The mechanism of geometric inversion may be explained on the basis of the intermediates 49 and 50, which allow rotation [41] of the —CHCl—CH₃ group about the single bond in compound 49 to the position appropriate for the chlorine to interact with iron to form the chloro carbonyl 51. This rotation results in the formation of the syn-1-methylallylchloroiron carbonyl 51.

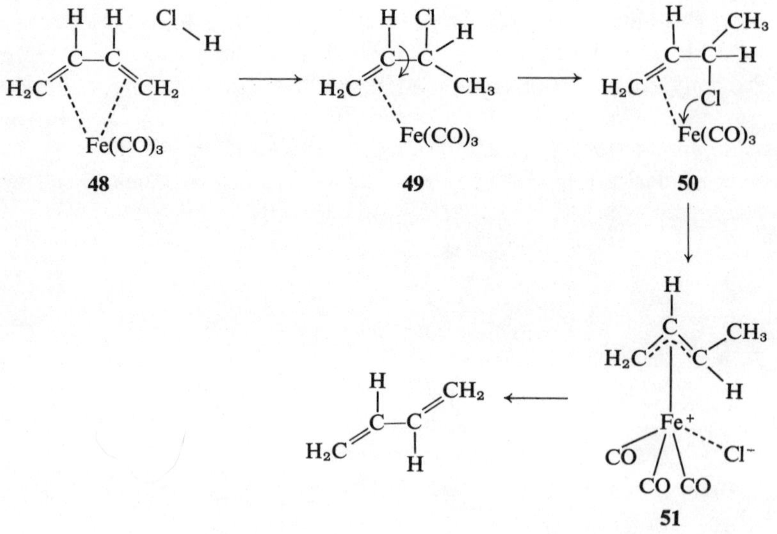

$$48 \longrightarrow 49 \longrightarrow 50$$

$$\text{(23)}$$

$$51$$

One of the simplest examples of geometric inversion is the formation of trans-1,3-pentadiene–iron tricarbonyl (53) from cis-1,3-pentadiene, 52. It is not possible to prepare cis-1,3-pentadiene–iron tricarbonyl by a direct reaction. The compound was indirectly obtained [44] by protonation of trans-2,4-pentadiene-1-ol-iron tricarbonyl 54 with strong acids, such as $HClO_4$, HBF_4, and HPF_6. The cation 55 has a cis configuration as confirmed by its NMR spectra. Geometric inversion has obviously accompanied the formation of 55 from 54. Reduction of 55 with sodium borohydride gives a mixture of 80% cis-1,3-pentadiene–iron tricarbonyl 56 and 20% of the trans isomer 57.

The geometric inversion accompanying the formation of the pentadiene–iron tricarbonyl cation 55 is similar to the protonation [45] of the

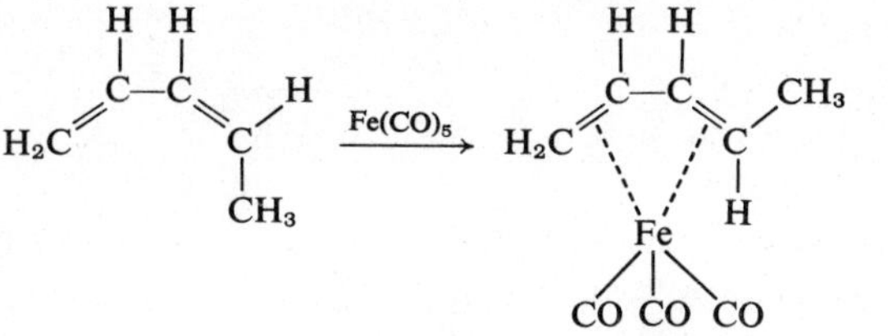

$$\tag{24}$$

52   **53**

54   **55**

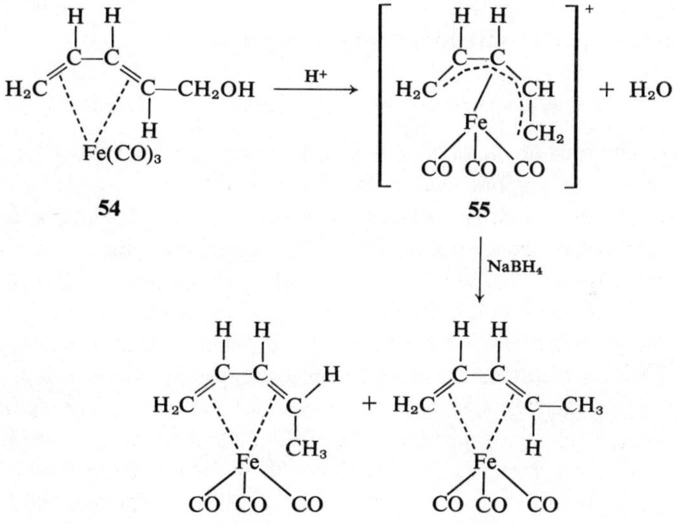

NaBH$_4$

$$\tag{25}$$

56   **57**

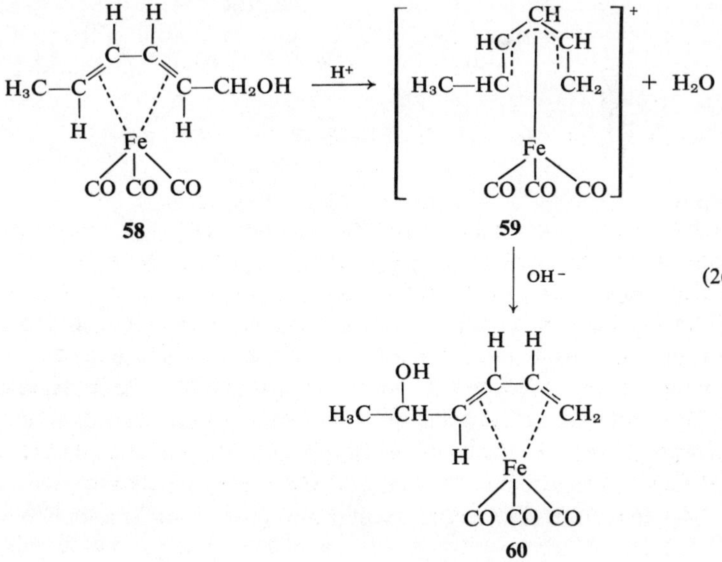

58   **59**

OH$^-$ $$\tag{26}$$

60

trans-2,4-hexadien-1-ol-iron tricarbonyl complex **58**, to yield *syn*-1-methyl-pentadienyliron tricarbonyl cation **59**. The cation formed reacts with water (i.e., hydroxide ion) to give exclusively an isomeric, *trans*-3,5-hexadiene-2-ol-iron tricarbonyl complex **60** without forming any of the original complex, **58**. It was proposed by Mahler *et al*. [45] that solvation of **59** gives an intermediate which is then converted to the diene complex, **60**, which is less hindered than that of the original diene complex, **58**.

II. Metal Complex-Catalyzed π-Bond Migration

A. CATALYSIS BY TRANSITION METAL COMPLEXES

Complex ions of the platinum group metals that are effective in isomerization of olefins include those of platinum(II), palladium(II), rhodium(I), iridium(I), and cobalt(I). Palladium(II) is by far the most widely studied metal ion for isomerization reactions, probably because of the similarity of the homogeneous reactions of its π-bonded complexes to the corresponding structures on the metal surface encountered in heterogeneous catalysis. The catalytic homogeneous reactions of platinum may thus be useful for elucidating the mechanisms of heterogeneous reactions, which are generally less amenable to mechanistic studies because of lack of microscopic information about catalytic metal surfaces. According to Rooney and Webb [47] the surface atoms in metals seem to possess chemical properties which are nearly identical to those of the free ions in many types of reactions.

The complex ions investigated by Harrod and Chalk [48] as catalysts for the isomerization of olefins include dichlorobis(ethylene)-μ,μ-dichlorodiplatinum(II), chlorobis(ethylene)dichlorodiplatinum(II), palladium(II) chloride, bis(benzonitrile)-μ,μ-dichloropalladium(II), rhodium(III) heptanoate, tetrakis(ethylene)-μ,μ-dichlorodirhodium(I), rhodium(III) chloride, and iridium(III) chloride. The trichlorides of rhodium and iridium are probably reduced to monovalent complexes in the presence of solvents such as mixtures of ethanol, isopropanol, or *t*-butanol with acetic acid. Rhodium trichloride, rhodium heptanoate, and iridium trichloride are effective as catalysts only in the presence of one of these solvent mixtures.

The equilibrium distribution of isomers obtained is sensitive to the nature of the metal complex used as catalyst, the oxidation state of the metal, and the ligands present on the metal ion. The isomerization reactions of 1-hexene, 2-heptene, and 3-heptene were studied [48] at 100°C in the presence of various catalysts and of various solvents as co-catalysts. The relative activities of the solvents decrease in the order, $C_2H_5OH > (CH_3)_2CHOH > (CH_3)_3COH \simeq CH_3COOH$. Water and hydrochloric acid were found to inhibit the isomerization reactions. The products of isomerization of 1-hexene were mainly cis and trans-2-hexene in the ratio 22:60 together with considerable trans-3-hexene. In the reactions catalyzed by platinum(II), palladium(II), and rhodium(I), an initial build-up of the cis-2-hexene isomer to more than the final equilibrium value (22%) was observed [48]. Rhodium(I) showed high selectivity favoring the formation of the cis isomer in the initial phases of the reaction, but the concentration of the cis isomer reached the equilibrium value at the end of the reaction. Preferential formation of the cis-2-isomer in the early phases of the isomerization reaction was also noticed by Bond and Hellier [49,50] in the isomerization of 1-pentene and by Sparke and Turner [51–53] in the isomerization of 4-methyl-1-pentene with bis(benzonitrile)–Pd(II) as the catalyst. The formation of the cis isomer in super-equilibrium amounts in the early phases of the isomerization reaction may be due to a higher stability of the metal cis isomer π-complex as compared to the trans-π-complex. Near the end of the reaction the ratio of cis/trans isomers is shifted in the direction of the more stable metal-free trans isomer.

Two mechanisms have been proposed for the metal ion-catalyzed isomerization: metal hydride addition and elimination, and a π-allyl mechanism.

In the metal hydride addition–elimination mechanism, addition of a pre-formed metal hydride to a coordinated olefin is considered to take place, with the formation of an alkyl σ complex. This is followed by re-formation of a metal hydride–olefin complex accompanied by stepwise migration of the double bond [Eq. (27)].

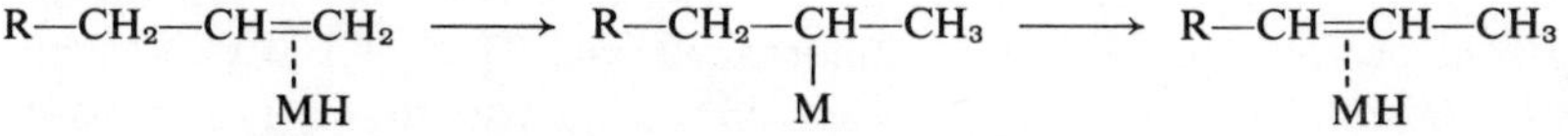

$$R-CH_2-CH{=}CH_2 \longrightarrow R-CH_2-CH-CH_3 \longrightarrow R-CH{=}CH-CH_3$$
$$\quad\quad\quad \overset{|}{MH} \quad\quad\quad\quad\quad \overset{|}{M} \quad\quad\quad\quad\quad\quad \overset{|}{MH} \quad\quad (27)$$

The π-allyl mechanism involves abstraction of a hydride ion from the coordinated olefin, followed by addition of the hydride at a carbon atom different from its original location:

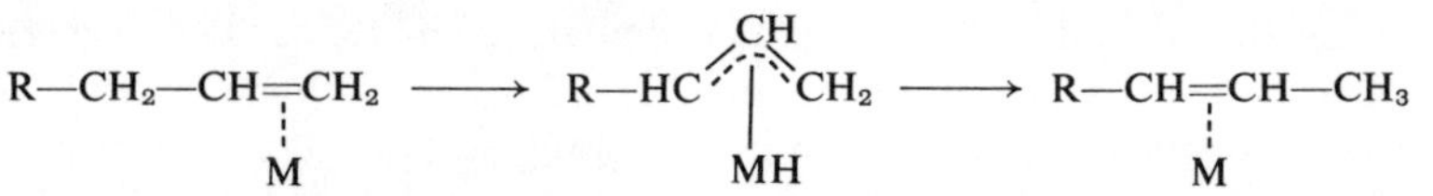

$$R-CH_2-CH{=}CH_2 \longrightarrow R-HC\overset{CH}{\underset{MH}{\diagup\diagdown}}CH_2 \longrightarrow R-CH{=}CH-CH_3 \quad (28)$$
$$\quad\quad\quad \overset{|}{M} \quad\quad\quad\quad\quad\quad\quad\quad\quad\quad\quad\quad\quad\quad\quad \overset{|}{M}$$

The metal hydride addition–elimination or alkyl reversal mechanism requires initial formation of a metal hydride to give the π-olefin complex shown above. According to Harrod and Chalk [48] the hydride ion for the formation of the metal hydride is either derived from a co-catalyst base (ethyl alcohol, acetone, or oxygen-containing organic compounds) or from the reaction of the carbanion formed by the initial nucleophilic attack of the co-catalyst anion on the coordinated olefin, followed by the release of a hydride ion. Coordinated olefins are very susceptible to such nucleophilic attack, as has been shown by Stern [54,55].

The protons derived from the acid co-catalyst or molecular hydrogen can also act as the source of hydride ion [56]. In the case of platinum(II), rhodium-(I), or iridium(I), direct formation of the metal hydride from the co-catalyst is possible because the metal ions in these cases can undergo oxidative addition of hydrogen or hydrogen ion.

$$M^{n+} + H^+ \longrightarrow HM^{(n+1)+} \tag{29}$$

$$M^{n+} + H_2 \longrightarrow H_2M^{n+} \tag{30}$$

In the case of palladium(II), it is highly unlikely that hydride is formed from the co-catalyst by two-electron oxidation–reduction reactions because of the very high potential of the palladium(II)–palladium(IV) oxidation. A possible source of the hydride ion for palladium(II) is the exchange of the vinylic hydrogen of the coordinated olefin with a halide ion coordinated to palladium, as shown in Eq. (31). An example of the vinylic hydrogen exchange depicted by reaction (31) is the isomerization of 1-butene with a

$$\overset{\diagdown}{\underset{\diagup}{\overset{\displaystyle CH}{\underset{\displaystyle CH}{\|}}}} \cdots Pd^+ \cdots Cl^- \qquad \overset{\diagdown}{\underset{\diagup}{\overset{\displaystyle CCl}{\underset{\displaystyle CH}{\|}}}} \cdots Pd\!-\!H \tag{31}$$

palladium(II) catalyst in the presence of C_2D_4, whereby deuterium is found to be incorporated in the isomerized 2-butene (deuterium exchange between C_2D_4 and 1-butene in the absence of metal is extremely slow). The use of deuterated acetic acid [57] and trifluoroacetic acid [56], CH_3COOD and CF_3COOD, in the isomerization of α-olefins with palladium(II) results in very little incorporation of deuterium in the isomerized olefin. Thus the co-catalyst is excluded as a source of hydride ion at least in the case of palladium(II).

Harrod and Chalk [58] have investigated in detail the steps involved in the migration of the π bond in vinyl- and allyl-deuterated α-olefins. The isomerization of 1-pentene-$1d_1$, $2d_1$ to 2-pentenes-$1d_{1.5}$, $2d_{0.5}$ with a metal hydride as catalyst may be summarized as given in Eq. (32).

Equilibria K_1 and K_2 allow exchange of deuterium at carbon atoms 1 and 2, whereas stepwise shifting of the double bond takes place by steps k_3 and k_4. The reaction with free olefin in step k_5 results in displacement of the 2-olefin formed in the first stage of the reaction, before further isomerization occurs. The observed fractional distribution of the label in the 2-pentene may be accounted for by the overall reactions (33) and (34). In step (33) migration of the double bond takes place without exchange of label at carbon atom 2, and in (34) with complete exchange of the label. The 50% exchange of label at carbon 2 observed in the reaction is a consequence of the ratio of reaction rate constants, $k_3'/k_4' = 1$. The movement of deuterium at each step of the

$$\begin{array}{c}
\underset{1}{}CHD \\
\|\text{------}MH \\
\underset{2}{}CD \\
| \\
(CH_2)_2 \\
| \\
CH_3
\end{array}
\underset{\xrightarrow{\ K_1\ }}{\rightleftharpoons}
\begin{array}{c}
CHD \\
|\diagdown M \\
CDH \\
| \\
(CH_2)_2 \\
| \\
CH_3
\end{array}
\underset{\xrightarrow{\ K_2\ }}{\rightleftharpoons}
\begin{array}{c}
CHD \\
\|\text{----}MH_{0.5}D_{0.5} \\
CD_{0.5}H_{0.5} \\
| \\
(CH_2)_2 \\
| \\
CH_3
\end{array}
\xrightarrow{\ k_3\ }
\begin{array}{c}
CH_{1.5}D_{1.5} \\
| \\
CD_{0.5}H_{0.5} \\
| \\
(CH_2)_2 \diagdown M \\
| \\
CH_3
\end{array}$$

$$\downarrow k_4 \qquad\qquad (32)$$

$$\begin{array}{c}
CH_{1.5}D_{1.5} \\
| \\
CD_{0.5}H_{0.5} \\
\| \\
CH \\
| \\
CH_2 \\
| \\
CH_3
\end{array}
+
\begin{array}{c}
CHD \\
\|\text{----}MH \\
CD \\
| \\
(CH_2)_2 \\
| \\
CH_3
\end{array}
\underset{\xleftarrow{\ \ k_5\ \ }}{+\ CHD{=}CD{-}(CH_2)_2{-}CH_3}
\begin{array}{c}
CH_{1.5}D_{1.5} \\
| \\
CD_{0.5}H_{0.5} \\
\|\text{------------}MH \\
CH \\
| \\
CH_2 \\
| \\
CH_3
\end{array}$$

$$CH_3CH_2CH_2CD{=}CDH \xrightarrow{\ k_3'\ } CH_3CH_2CH{=}CD{-}CH_2D \qquad (33)$$

$$CH_3CH_2CH_2CD{=}CDH \xrightarrow{\ k_4'\ } CH_3CH_2CH{=}CH{-}CD_2H \qquad (34)$$

reaction has been followed by NMR spectrometric measurement of the vinylic and allylic protons. The observed isotope distribution supports the alkyl reversal mechanism for stepwise migration of the double bond, as indicated in Eq. (32), since the π-allyl mechanism would also place some label on carbon atom 3.

In the case of allyl deuterated α-olefins [58], such as 1-heptene-3d_2 the label was spread to both carbon atoms 1 and 2, indicative of the overall steps (35) and (36). The observed distribution [58] of the label at carbon atoms 1, 2,

$$C_4H_9CD_2CH{=}CH_2 \xrightarrow{\ k_5\ } C_4H_9CD{=}CH{-}CDH_2 \qquad (35)$$

$$C_4H_9CD_2CH{=}CH_2 \xrightarrow{\ k_6\ } C_4H_9CD{=}CD{-}CH_3 \qquad (36)$$

and 3 in the final product, $C_4H_9CD{=}CH_{0.5}D_{0.5}{-}CH_{2.5}D_{0.5}$ corresponds to the ratio $k_5/k_6 = 1$, in accord with the alkyl reversal mechanism.

In the alkyl reversal mechanism of Harrod and Chalk [48] [Equation (32)] the rates of formation of various isomers depend on the rate of isomerization of the olefin coordinated to the metal ion and the rate of exchange of free and coordinated olefin. In the initial phases of the reaction, cis-2-pentene is formed in super-equilibrium amounts probably due to the displacement step k_5 being comparable to or faster than the conversion of cis- to trans-olefins. This is due to the fact that the cis-olefin is more stable in the form of the metal π complex, while the trans form is the more stable form for the metal-free olefin.

Cramer and Lindsey [56] and Cramer [59] have studied the isomerization of α-olefins in the presence of several catalysts: rhodium(I) + H^+ (proton from acid or solvent, CH_3OH); $[(C_2H_4)_2PtCl_2]_2$ + H^+; H_2PtCl_6 + $SnCl_2$ + H_2 (source of H^+); H_2PtCl_6 + $SnCl_2$ + methanolic HCl; $[(C_6H_5O)_3P]_4Ni(0)$ + H^+; and $H_2Fe(CO)_4$ + H^+, and have suggested a modified hydride addition–elimination reaction that involves the following steps:

(a) Oxidative addition of H^+(HCl) or molecular hydrogen to the metal complex

(b) Olefin exchange of the hydrido metal complexes with the original olefin

(c) Hydride addition to the complexed olefin yielding a σ-bonded alkyl complex

(d) Olefin exchange of the isomerized olefin complex

The isomerization of 1-butene with rhodium(I), platinum(II), nickel(0), and iron(−2) catalysts in CH_3OD in the temperature range −25° to 0°C indicates the following:

(1) The cis- and trans-2-butenes formed are mostly nondeuterated and all the label is confined to 1-butene at carbon 2 as confirmed by NMR and IR studies.

(2) One deuterium is introduced for each molecule of 1-butene that is isomerized to cis- and trans-2-butenes.

(3) The rate of isomerization/deuteration varies with the catalyst. For rhodium(I), the ratio is approximately 1:1 whereas for platinum(II), nickel(0), and iron(−2) [e.g., $H_2Fe(CO)_4$] catalysts, the rate of isomerization is much faster than that of deuteration.

Observations 1–3 may be explained by reaction scheme (37).

The metal ion in the olefin π complex **61** is oxidized by D^+ (or H^+) in step k_1 to form the hydrido complex **62** which on rearrangement forms the σ complex **63**. Complex **63** then rearranges to the hydridometal–olefin complex **64**, which undergoes an olefin displacement reaction with 1-butene to release 2-deutero-1-butene, $CH_2\!\!=\!\!CDC_2H_5$, in solution. The step corresponding to k_4 thus accounts for the formation of 1-butene-2d. Complex **65** undergoes hydride addition to form the alkyl complex **66** that isomerizes to 2-butene complex **67**. Removal of H^+ from **67** results in the formation of the M^{n+}–2-butene complex **68** that undergoes olefin exchange with 1-butene to form **61** and nondeuterated 2-butenes.

The difference in the rates of isomerization and deuteration for various catalysts [rhodium(I), platinum(II), nickel(0), and $H_2Fe(CO)_4$] depends on the relative tendencies of complexes such as **66** and **64** to undergo isomerization and exchange. This may be dependent on the nature of the metal ion

$$\underset{\textbf{61}}{\overset{M^{n+}}{\underset{|}{CH_2\!=\!CHC_2H_5}}} \xrightarrow[k_1]{D^+} \underset{\textbf{62}}{\overset{DM^{(n+1)+}}{\underset{|}{CH_2\!=\!CHC_2H_5}}} \xrightarrow[k_2]{}$$

$$\underset{\textbf{63}}{\overset{M^{(n+1)+}}{\underset{|}{CH_2\!-\!CHD\!-\!C_2H_5}}} \xrightarrow[k_3]{} \underset{\textbf{64}}{\overset{HM^{(n+1)+}}{\underset{|}{CH_2\!=\!CDC_2H_5}}}$$

$$k_4 \Big| + CH_2\!=\!CH\!-\!C_2H_5$$

$$\underset{\textbf{65}}{\overset{HM^{(n+1)+}}{CH_2\!=\!CDC_2H_5 \ + \ \underset{|}{CH_2\!=\!CH\!-\!C_2H_5}}}$$

$$k_5 \Big|$$

$$k_8 \quad \begin{array}{c} +(CH_2\!=\!CH\!-\!C_2H_5) \\ -(CH_3CH\!=\!CH\!-\!CH_3) \end{array}$$

$$\underset{\textbf{68}}{\overset{M^{n+}}{\underset{|}{CH_3CH\!=\!CH\!-\!CH_3}}} \xleftarrow[k_7]{-H^+} \underset{\textbf{67}}{\overset{HM^{(n+1)+}}{\underset{|}{CH_3CH\!=\!CHCH_3}}} \xleftarrow{k_6} \underset{\textbf{66}}{\overset{M^{n+}}{\underset{|}{CH_3CHC_2H_5}}} \qquad (37)$$

bound to the olefin and the relative stabilities of the end and internal olefin–metal complexes. Although most of the deuterium is incorporated in 1-butene, a small amount of the label goes into the isomerized 2-butenes by hydride addition–elimination reaction with migration of the π bond in complex **64** to form $CH_3CD\!=\!CHCH_3$ (i.e., reaction sequence k_5, k_6 occurs prior to exchange, k_4). It is also possible to eliminate a proton from **65** or a deuteron from **62** to form complex **61** without any isomerization. In general if the $\sigma1$ complex **63** is more stable than the $\sigma2$ complex **66**, and the ligand (olefin) exchange reaction is rapid, the main deuterated product will be 2-deutero-1-butene, even though isomerism to 2-butene requires the formation of the $\sigma2$ complex as an intermediate.

The above observations are in accord with the hydride addition–elimination mechanism mainly because (1) in the π-allyl mechanism no exchange is possible with the solvent and (2) the formation of non-deuterated 2-butenes and C-2–deuterated 1-butene cannot be accounted for in any other way.

cis–trans Isomerism in these reactions is highly dependent on the catalyst. Thus the ratio of *trans* to *cis* butenes in the case of rhodium(I) catalysis was about unity, but in the case of platinum(II), nickel(0), and $H_2Fe(CO)_4$ the ratio was greater than 3:1.

Milogram and Urey [60] described the isomerization of 1-hexene and 1-octene with olefin–Pt(II) complexes as catalysts in the presence of hydrochloric acid (0.03 M). Formation of a σ-bonded cationic intermediate (**70**) from the olefin complex **69** was postulated. Loss of a proton then would give

71 and the rearranged olefin is then displaced from the metal ion by the original olefin. Although metal hydrido groups are not suggested as intermediates in these reactions, the active Pt(II)–σ-bonded intermediate illus-

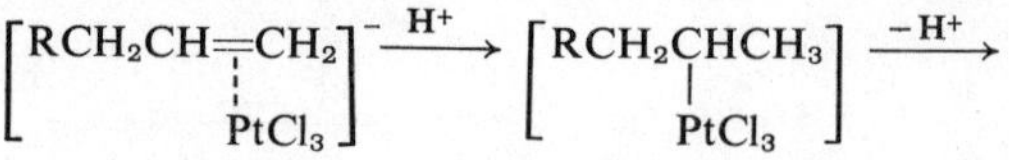

$$\left[\begin{array}{c} RCH_2CH{=}CH_2 \\ | \\ PtCl_3 \end{array}\right]^- \xrightarrow{H^+} \left[\begin{array}{c} RCH_2CHCH_3 \\ | \\ PtCl_3 \end{array}\right] \xrightarrow{-H^+}$$

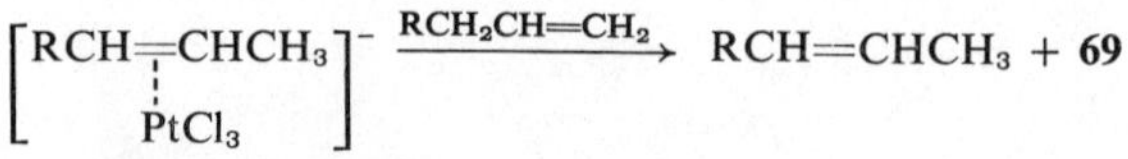

$$\underset{69}{} \qquad\qquad \underset{70}{}$$

$$\left[\begin{array}{c} RCH{=}CHCH_3 \\ | \\ PtCl_3 \end{array}\right]^- \xrightarrow{RCH_2CH=CH_2} RCH{=}CHCH_3 + 69$$

$$\underset{71}{}$$

trated by formula 70 is similar to the σ-bonded intermediate in the alkyl reversal mechanism of Harrod and Chalk [48].

In an interesting example of the isomerization of a conjugated to a nonconjugated diene described by Rinehart and Lasky [43], 1,3-cyclooctadiene, 72, is rearranged to the nonconjugated 1,5-isomer through the formation of an intermediate rhodium(I) complex. A dimer 73 of 1,5-cyclooctadienechlororhodium(I) is initially formed from which 1,5-cyclooctadiene may be obtained by displacement with cyanide ion in 66% yield. Although 1,3-cyclooctadiene is by itself thermodynamically more stable than the 1,5-isomer, the driving force for the reaction seems to be the greater chelate stability of the 1,5-isomer.

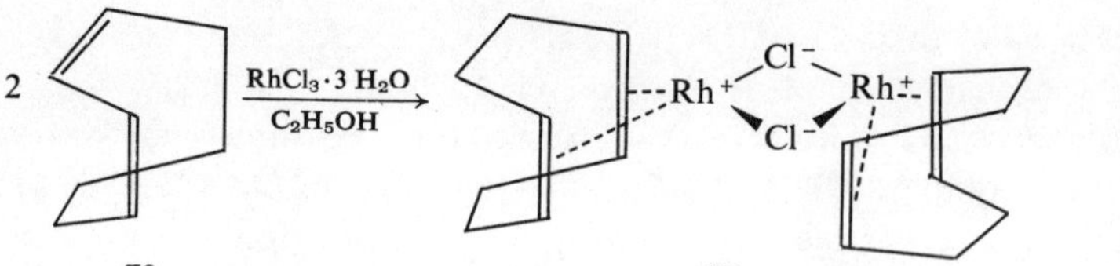

$$(38)$$

$$\underset{72}{} \qquad\qquad \underset{73}{}$$

The mechanism proposed for the isomerization of the 1,3-cyclooctadiene to the 1,5-isomer is initial formation of a Rh(I) π complex 74 [Eq. (39)] which undergoes an intramolecular hydride abstraction to form a π-allyl hydride intermediate 75 followed by rearrangement to the 1,5-isomer [43]. The formation of the 1,5-isomer may proceed through an intermediate 1,4-isomer which could not be detected in the reaction.

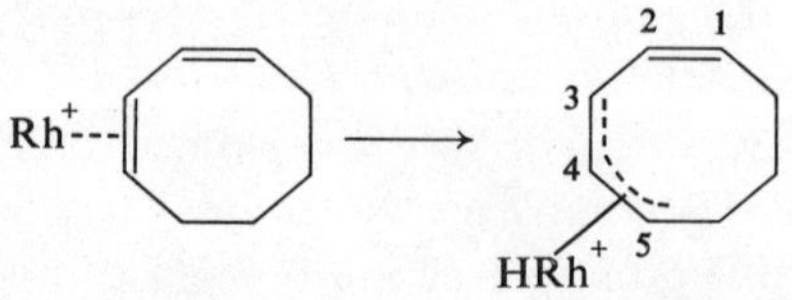

$$(39)$$

$$\underset{74}{} \qquad\qquad \underset{75}{}$$

The balance between thermodynamic stability of a diene and the stability of its metal π complex is reflected in the isomerization of 1,5-cyclooctadiene to the 1,3-isomer in the presence of bis(triphenylphosphine)dichloroplatinum-(II) and stannous chloride as catalysts [61]. Stannous chloride seems to be essential for the isomerization reaction. The effective reaction intermediate in the isomerization may be a hydridoplatinum–olefin complex PtH(SnCl$_3$)-(olefin)(PPh$_3$)$_2$ which has been isolated from the reaction mixture and identified. The kinetics suggests that isomerization occurs in two consecutive reversible first-order reactions by a hydride addition–elimination mechanism [61], indicated in reaction sequence (40).

$$1,5\text{-Cyclooctadiene} \underset{k_2}{\overset{k_1}{\rightleftarrows}} 1,4\text{-cyclooctadiene} \underset{k_4}{\overset{k_3}{\rightleftarrows}} 1,3\text{-cyclooctadiene} \quad (40)$$

Migration of the double bond, which has not been observed for the isomerization of 1,5-cyclooctadiene with rhodium(I) catalysts, has been found experimentally with the platinum(II) catalysts, possibly because of the greater stabilities of the intermediate isomeric complexes of platinum(II).

Another example of migration of the π bond in a cyclic system is the rhodium(I)-catalyzed isomerization of *cis,trans*-1,5-isomer of cyclodecadiene, **76**, to *cis,cis*-1,6-cyclodecadiene with formation of a dimeric 1,6-cyclodecadiene–Rh(I) chloride complex, **77** [62] [Eq. (41)]. The oxidation of ethanol in

$$2 \quad \overset{RhCl_3}{\underset{EtOH}{\longrightarrow}} \quad \text{(complex)} \quad (41)$$

76   **77**

reaction (41) indicates that the formation of the 1,6-cyclodecadiene–rhodium-(I) chloride dimer may involve reduction of rhodium(III) by a hydrogen abstraction mechanism [Eq. (42)]. The mechanism of isomerization of the

$$C_2H_5OH + RhCl_3 \longrightarrow HRhCl_2 + HCl + CH_3CHO \quad (42)$$

1,5-diene to the 1,6-isomer is not quite clear and may involve a process similar to Eq. (39), or reversible hydride addition and elimination, since hydride ion is available by interaction of the solvent with rhodium trichloride.

Bond and Hellier [49,50] studied the isomerization of 1-pentene with palladium(II) chloride or the bis(benzonitrile)-Pd(II) complex as catalyst. Benzene, ethyl acetate, and methyl ethyl ketone were found to be suitable solvents for the reaction, which proceeds very slowly in the absence of these solvents. A π-allyl mechanism, indicated in reaction sequence (43), was proposed for the isomerization reaction. Palladium(II) forms very well defined

π-allyl complexes in solution, the structures of which are now substantiated [63] by X-ray studies. According to Bond and Hellier [49,50], a palladium(II)–olefin complex **78** is first formed from palladium(II) chloride and 1-pentene. This complex is very labile [64] at 70°C and is believed to form the π-allyl intermediate **79** quite readily. Return of the hydride ions in **79** to carbon atoms other than those to which they were originally bonded yields the co-ordinated isomerized olefin **80**, which is displaced by the original olefin, as indicated in Eq. (43). This reaction results in the initial formation of *cis*-2-pentene from 1-pentene in concentrations higher than that predicted by the uncatalyzed equilibrium between the two isomers.

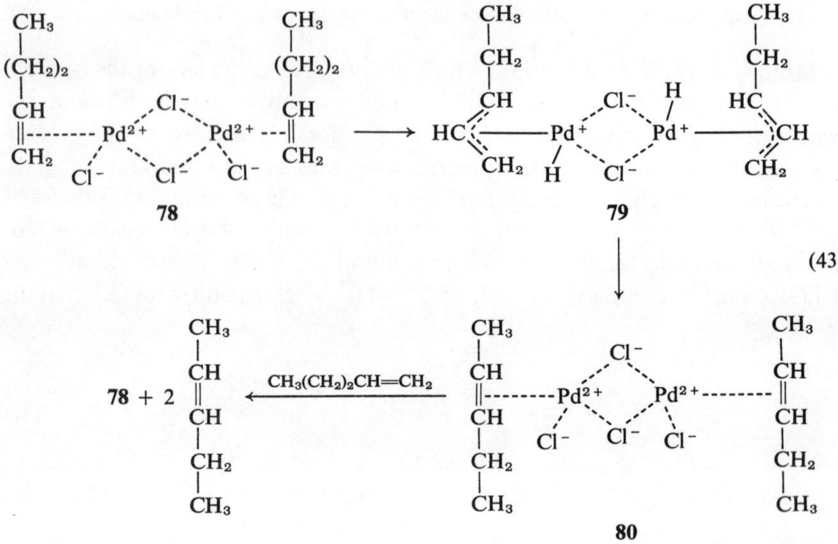

(43)

An objection to the π-allyl mechanism has been the inability of preformed palladium–π-allyl complexes to catalyze isomerization of olefins under mild conditions, but it is suggested that a labile π-allyl complex of palladium(II) such as **79**, is formed *in situ* and is an intermediate in isomerization reactions. The case is similar to the inability of some of the stable hydrido complexes such as $PtHCl(PPh_3)_2$ to catalyze isomerization of olefins under mild conditions [51], although labile metal hydrides have been proposed as intermediates in platinum complex-catalyzed isomerization reactions.

cis-2-Pentene was isomerized to *trans*-2-pentene to the extent predicted from the relative thermodynamic stabilities of these *cis* and *trans* isomers. The mechanism suggested [49,50] for this isomerization involves intermediate formation of 1-pentene, which acts as a catalyst for the reaction. The rate of isomerization of *cis*-2-pentene was accordingly found to be very much slower than that of 1-pentene. There is an induction period [49] in the isomerization of the *cis*-2 isomer during which 1-pentene achieves its equilibrium concentration of about 5%. This induction period is removed on the addition of 1-pentene to the reaction mixture, thus indicating the importance of 1-pentene as an intermediate in the *cis–trans* isomerization of *cis*-2-pentene.

Rooney and Webb [47] have postulated 1-butene as an intermediate in the heterogeneous *cis–trans* isomerization of *cis*-2-butene. The mechanism which they suggested is presented in the reaction scheme (44). The order in which the

$$
\begin{array}{ccc}
\underset{\textit{trans-2-Butene}}{\overset{\displaystyle H_3C\diagdown\;\diagup H}{\underset{\displaystyle H\diagup\;\diagdown CH_3}{C=C}}}
\rightleftharpoons
\underset{\displaystyle H-M}{\overset{\displaystyle CH\quad CH_3}{\overset{\displaystyle H_2C\diagup|\diagdown CH}{}}}
& & \\[2em]
& & CH_2\!=\!CHCH_2CH_3 \rightleftharpoons \text{1-Butene} \\[1em]
\underset{\textit{cis-2-Butene}}{\overset{\displaystyle H_3C\diagdown\;\diagup CH_3}{\underset{\displaystyle H\diagup\;\diagdown H}{C=C}}}
\rightleftharpoons
\underset{\displaystyle H-M\;\;CH_3}{\overset{\displaystyle CH}{\overset{\displaystyle H_2C\diagup|\diagdown CH}{}}}
& & M
\end{array}
\tag{44}
$$

transition metals preferentially form π-allyl complexes is iron > palladium > cobalt > nickel $\simeq$ platinum > rhodium. Thus the feasibility of catalysis through the π-allyl mechanism in palladium(II) complexes would seem to be much greater than for platinum(II) or rhodium(I).

Davies [57,65] used tetrachloro-μ,μ-dichlorodipalladate(II) as a catalyst in the isomerization of 1-octene at 65°C. The products of isomerization were analyzed by vapor-phase chromatography and NMR spectra. Since on using CH_3COOD along with the palladium(I) catalyst it was found [57] that deuterium is not incorporated in the isomerized olefin, it was concluded that hydrogen is not lost from the substrate hydrocarbon at any stage, and the mechanism thus seems to be intramolecular without participation of the solvent. The formation of palladium hydride intermediates in palladium(II) catalysis was questioned [57] on the basis of nondetectability of such intermediates by NMR. It is nevertheless possible that labile hydride intermediates in question may have only transitory existence and may very well escape detection by magnetic resonance measurements.

In order to test the validity of the π-allyl mechanism in the isomerization of olefins by palladium(II) catalysis, Davies synthesized 3,3-dideutero-1-octene, $C_5H_{11}CD_2CH{=}CH_2$, and studied the products of its isomerization by vapor-phase chromatography. It was found [65] that deuterium does not become attached to the terminal carbon atom in the isomerized olefin. Davies [65] pointed out that if a π-allyl mechanism is operative in the isomerization of 3-deuterated 1-octene (**81**) then a 1,3-hydride shift through a π-allyl intermediate **82** should give rise to the end-deuterated olefin as in **83**. Since no deuterium could be detected [65] at the 1-position in the isomerized olefin, it was suggested that the olefin may pass through a coordinated carbene stage

$$-CD_2-CH{=}CH_2 \longrightarrow -DC\overset{\overset{\displaystyle CH}{\diagup\,\diagdown}}{\underset{\displaystyle D-Pd}{}}CH_2 \longrightarrow -CD{=}CH-CH_2D \quad (45)$$

$$\underset{\displaystyle Pd}{} \qquad\qquad\qquad\qquad \underset{\displaystyle Pd}{}$$

81   82   83

85 rather than a π-allylic complex **82**, with stepwise movement of hydrogen atoms from one carbon atom to the next as shown in reaction scheme (46).

$$RCD_2-\overset{\frown}{CH}{=}CH_2 \longrightarrow RCD_2-\overset{\frown}{C}-CH_3 \longrightarrow RCD{=}CD-CH_3 \quad (46)$$

$$\underset{\displaystyle Pd}{} \qquad\qquad \underset{\displaystyle Pd}{} \qquad\qquad \underset{\displaystyle Pd}{}$$

84   85   86

If such a mechanism is operative in the isomerization of olefins, substitution of hydrogen by methyl at carbon atom 2 in **84** should prevent such stepwise movement of hydrogen and thus block isomerization of a 2-substituted α-olefin. Davies [57,65] reported one such example of the inactivity of 2-methyl-1-pentene toward isomerization under the experimental conditions employed. Sparke and Turner [52,53], however, found that 2-methyl-1-pentene (**87**) is isomerized at 50°C under a nitrogen atmosphere by bis(benzonitrile)–Pd(II) chloride, the products of isomerization being 2-methyl-2-pentene (**88**), 4-methyl-1-pentene (**89**), and 4-methyl-2-pentene (**90**). The isomerization of

$$CH_2{=}\underset{\displaystyle \underset{\displaystyle CH_3}{|}}{C}-CH_2CH_2CH_3 \longrightarrow CH_3-\underset{\displaystyle \underset{\displaystyle CH_3}{|}}{C}{=}CHCH_2CH_3 \;+$$

87   88

75%

$$CH_3-\underset{\displaystyle \underset{\displaystyle CH_3}{|}}{CH}-CH{=}CHCH_3 \;+\; CH_3-\underset{\displaystyle \underset{\displaystyle CH_3}{|}}{CH}-CH_2-CH{=}CH_2 \quad (47)$$

90   89

0.5%   11.3%

2-methyl-1-pentene thus seems to favor a π-allyl mechanism over the carbene mechanism for the isomerization of α-olefins.

Harrod and Chalk [66] have provided evidence in favor of the reversible hydride addition–elimination equilibrium through a study of the isomerization of $R-CH_2-CD{=}CD_2$ and $R-CH_2CD{=}CDH$ by palladium(II) chloride. It was found [66] that the end-deuterated compounds isomerize three times faster than the corresponding protonated olefins. The rates of isomerization were followed by gas chromatography and infrared spectra. In accordance with the hydride addition–elimination mechanism two reactions, (48) and (49), are possible for the isomerization of end-deuterated olefins.

$$R—CH_2—CD{=\!=}CDH \longrightarrow R—CH_2—CD—CDH_2 \longrightarrow R—CH{=\!=}CD—CDH_2 \quad (48)$$
$$\phantom{R—CH_2—CD{=\!=}CDH \longrightarrow} MH M MH$$

$$R—CH_2—CD{=\!=}CDH \longrightarrow R—CH_2—CDH—CDH \longrightarrow R—CH_2—CH{=\!=}CDH \quad (49)$$
$$\phantom{R—CH_2—CD{=\!=}CDH \longrightarrow} MH M MD$$

In the first reaction, a C—H bond is broken with resulting isomerization of the olefin. Metal–hydride deuterium exchange takes place in the second reaction, (49), with no net isomerization.

The successive addition and elimination of the metal hydride in the above reactions raises the question of the hydride source for the formation of the palladium hydride. That solvents and co-catalysts are not the source of the hydride species results from a number of observations [56,65] in which no exchange takes place with solvents such as CH_3COOD and CF_3COOD. Thus vinylic hydrogen seems to be the only source of hydride for the formation of the intermediate hydride complexes. The intermolecular nature of the hydride transfer in the palladium(II)-catalyzed isomerization reaction is indicated by the isomerization of 1-pentene in the presence of 1-heptene-3d_2 [58] whereby the label is distributed over all the carbon atoms of the isomerized 2-pentenes.

In the case of allylic deuterated 1-olefins (with deuterium at the 3-position) a hydride addition–elimination mechanism [61] similar to that indicated in Eqs. (48) and (49) may be postulated as shown in Eqs. (50) and (51). The first

$$R—CD_2—CH{=\!=}CH_2 \longrightarrow R—CD_2—CH—CH_3 \longrightarrow R—CD{=\!=}CH—CH_3 \quad (50)$$
$$\phantom{R—CD_2—CH{=\!=}CH_2 \longrightarrow} MH M MD$$

$$R—CD_2—CH{=\!=}CH_2 \longrightarrow R—CD_2—CDH—CH_2 \longrightarrow R—CD_2—CD{=\!=}CH_2 \quad (51)$$
$$\phantom{R—CD_2—CH{=\!=}CH_2 \longrightarrow} MD M MH$$

reaction sequence involves the breaking of a C—D bond while the second involves breaking a C—H bond. If breaking of a C—D bond is slower than breaking a C—H bond, isomerization of a C3–deuterated olefin should be slower than that of an end-deuterated olefin. The recent observation of Harrod and Chalk [58] that the 3-deuterated 1-olefins isomerize at the same rate as the undeuterated 1-olefins casts some doubt on the above reaction scheme. More-over, when vinyl-deuterated 1-pentene, $CH_3CH_2CH_2CD{=\!=}CDH$, was isomerized with palladium(II), all the possible internal isomers appeared at comparable rates [58]. This observation is inconsistent with a hydride addi-tion–elimination mechanism where stepwise migration of the double bond results in the building up of the concentration of one of the isomers ahead of the formation of the next internal isomer. Although Harrod and Chalk [58] have proposed an intramolecular shift of hydrogen for palladium(II)-catalyzed reactions, the details of such a mechanism are far from clear. It is also possible that palladium(II)-catalyzed reactions take place by a combination of the π-allyl and hydride addition–elimination mechanism. Thus the π-allyl step would be the source of hydride and would be followed by the hydride addition–elimination step. Elucidation of the microscopic details of palladium(II)-catalyzed isomerization reactions required further work along these lines.

Sparke and Turner [51–53] favor a palladium(II)–olefin complex as catalyst in the isomerization of olefins, with the rearrangement taking place either within the complex or when the olefin leaves or enters the complex. The actual steps by which rearrangements occur were not suggested; however, it may be recalled that formation of π-olefin complexes is assumed in the first step of both the alkyl reversal mechanism of Harrod and Chalk [48,66] and of the π-allyl mechanism of Bond and Hellier [49,50].

Cruikshank and Davies [67] have studied the isomerization of allylbenzene to trans-methylstyrene in the presence of palladium(II) chloride in acetic acid. The isomerization reaction has been found to be first order with respect to the catalyst and zero order with respect to allylbenzene. The slow step in the reaction has been postulated to be the rate-determining conversion of the original palladium(II)–olefin complex to an intermediate hydrido species prior to isomerization.

Kovacs et al. [68] have studied the isomerization of 1-hexene in the presence of a catalyst consisting of dinitrogentris(triphenylphosphine)cobalt(0), $Co(N_2)(PPh_3)_3$, 91. 1-Hexene was isomerized to a mixture of 55% cis-2-hexene, 35% trans-2-hexene, and 10% 3-hexene. The rate of isomerization was found [68] to be second order with respect to a catalyst–1-hexene complex having the composition $(1\text{-hexene})(PPh_3)_3Co$. The rate law is given by Eq. (52).

$$-\frac{d[1\text{-hexene}]}{dt} = k[(1\text{-hexene})Co(PPh_3)_3]^2 \tag{52}$$

The mechanism of Eqs. (53)–(55) was proposed to explain these observations [68]. The value of the rate constant for reaction (54) was reported as 6.10 M^{-1} second^{-1} at 25°C. The activation parameters are $\Delta H^{\ddagger} = 10.1$ kcal/

$$Co(N_2)(PPh_3)_3 + 1\text{-hexene} \xrightarrow{\text{fast}} (1\text{-hexene})(PPh_3)_3Co + N_2 \tag{53}$$

$$(1\text{-hexene})(PPh_3)_3Co \xrightarrow{k} [(2\text{-hexene})(PPh_3)_3Co]_2 \tag{54}$$

$$[(2\text{-hexene})(PPh_3)_3Co]_2 + 2\,[1\text{-hexene}] \xrightleftharpoons{K} 2\,(1\text{-hexene})Co(PPh_3)_3 + 2\,[2\text{-hexene}] \tag{55}$$

mole and $\Delta S^{\ddagger} = -21$ eu. It was proposed that olefin isomerization takes place within binuclear complexes **92** and **93** by the following scheme:

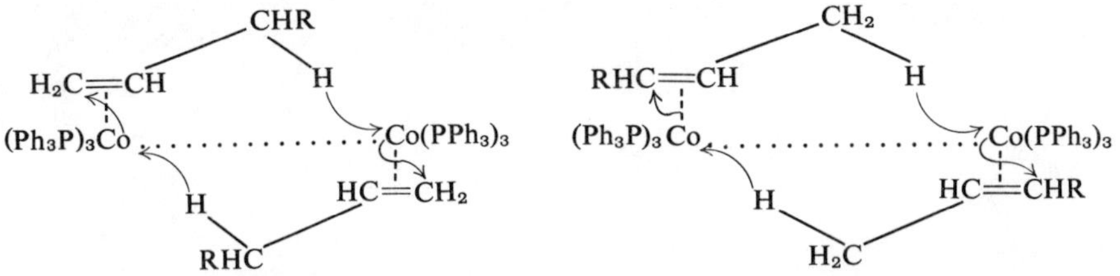

In addition to the "3,1'-hydride shift" illustrated by **92** and **93**, reasonable alternatives are the standard olefin isomerization mechanisms: (1) the addition and elimination of a metal hydride and (2) rearrangement through a π-allyl hydride.

B. ISOMERIZATION OF OLEFINS BY SILVER(I)

Kisker and Crandale [69] reported homogeneous silver ion-catalyzed isomerization of 4-maleylacetoacetic acid **(94)** and maleylacetone **(96)** to the corresponding *trans* isomers (fumaryl derivatives), as shown in Eqs. (56)

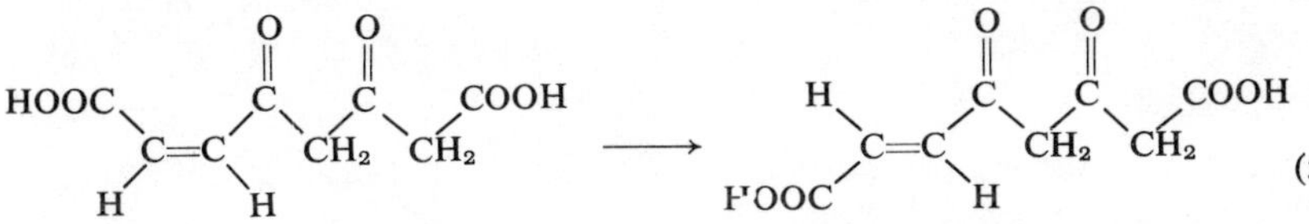

and (57). Silver ion is reported [69] to be specific for this reaction. The metal ions copper(I), mercury(II), aluminum(III), cadmium(II), cobalt(II), copper(II), chromium(III), iron(II), iron(III), mercury(I), magnesium(II), manganese(II), nickel(II), lead(II), platinum(IV) and zinc(II) were all found to be ineffective for the *cis–trans* isomerization of these maleyl derivatives. The isomerization with silver(I) ion was followed spectroscopically by measuring the increase in absorption at 315 nm of aliquots of reaction mixtures. No mechanism was suggested for this reaction, and the nature of the intermediate silver(I) complexes is open to conjecture.

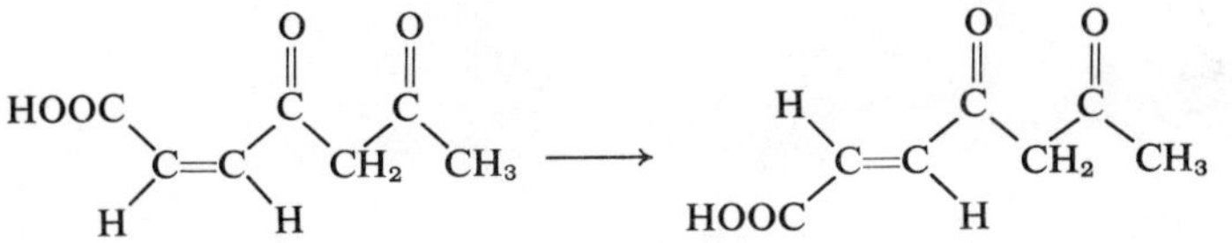

The Oxo Reaction

3

I. Introduction

The oxo reaction was developed by Otto Roelen [70–73] through modification of the Fischer-Tropsch synthesis in order to produce aldehydes and ketones rather than hydrocarbons as the main products. The reaction was named the "oxo" synthesis because it resulted in the formation of oxygenated products from hydrocarbons. It is also referred to as hydroformylation, since the net process involves the addition of hydrogen and formyl groups across the double bond of an alkene. The two terms "oxo" and "hydroformylation" refer to the same reaction and are used interchangeably in this book. The general "oxo" reaction may be represented as in Eq. (58).

$$RCH{=}CH_2 + CO + H_2 \xrightarrow{\text{metal carbonyl}} \begin{array}{c} RCH_2CH_2CHO \\ \diagup \\ \diagdown \\ RCH(CHO)CH_3 \end{array} \qquad (58)$$

The oxo reaction is catalyzed by a variety of metal carbonyls and other complexes, such as cobalt carbonyls [74], rhodium carbonyls [75], iron and manganese carbonyls [76], bis(triphenylphosphine)rhodium(I) [77,78], and hydridocarbonyltris(triphenylphosphine)rhodium(I) [79,80].

The primary products of hydroformylation of olefins are aldehydes, which can be converted readily to secondary products such as primary alcohols by subsequent hydrogenation reactions. The "oxo alcohols," which constitute the main commercial products of the oxo reaction, have found a host of applications [81] in industry. Several other intermediates and commercially attractive products may be obtained from aldehydes. The oxo reaction has thus commanded attention from both industrial and academic research groups. Excellent surveys of the oxo reaction covering the period from 1955 to 1968 have been published by Orchin [82], Goldfarb and Orchin [83], Wender *et al.* [15,74], Hurd and Gwynn [84], Zachry [76], Lemke [85], and Chalk and Harrod [86]. Much work has been done recently on the kinetics and mechanisms of oxo reactions and of various related reactions that take place under oxo conditions such as hydrogenation of aldehydes and isomerization of olefins. In the present section emphasis is given primarily to the kinetics and mechanistic aspects of hydroformylation, with special attention to the function of metal ion catalysis in the activation of π bonds of alkenes and alkynes. The bibliography of the oxo reaction compiled in this section pertains mainly to this objective; nevertheless, it is extensive and covers the broad area of oxo research up to the date of compilation of this book.

II. Reactions of Alkenes

A. Conditions of the Reaction

The reaction conditions normally employed for the oxo reaction vary over the temperature and pressure ranges of 50°–200°C and 100–400 atm. A 1:1 molar ratio of carbon monoxide and hydrogen (synthesis gas, water gas) is usually employed. These wide variations of temperature and pressure permit considerable variation in product distribution. Thus a drop of temperature from 90° to 40°C reduces the rate [87,88] of hydroformylation of 1-butene by a factor of 100 at a pressure of 3000 lb/inch^2 of synthesis gas. Maintaining temperature below 120°C favors the formation of aldehydes, and at higher temperatures (about 180°C) alcohols are the main products.

The distribution of branched-chain and straight-chain isomers of aldehydes and alcohols is influenced by varying both temperature and pressure. Low temperature and higher pressure favor the formation of straight-chain isomers [88–90], whereas higher temperatures and low pressure [88] (less than 70 atm) yields a high percentage of branched-chain isomers. At pressures of less than 40 atm of synthesis gas, the metal carbonyl catalyst is unstable [88] and it decomposes at and above 150°C. The influence of temperature and pressure on product distribution, and implications concerning reaction mechanisms of hydroformylation, will be discussed in some detail in Section II,C.

Cobalt carbonyls are the most commonly employed homogeneous catalysts in hydroformylation of alkenes. Cobaltous compounds may be used in place of cobalt carbonyls but the reaction then requires temperatures higher than

120°C, in order to reduce the cobaltous compounds to the carbonyls [74] required for oxo catalysis. If the reaction is to be conducted below or at 120°C, direct addition of cobalt carbonyls to the reaction mixture is usually preferred. Cobalt carbonyl species present under oxo conditions are dicobalt octacarbonyl and cobalt hydrocarbonyl. The former, $Co_2(CO)_8$, is rapidly converted at 100°C by hydrogen and carbon monoxide under pressure to the latter, $HCo(CO)_4$. The hydrocarbonyl may be isolated as an anion in the absence of olefin by rapidly cooling the pressure vessel to $-50°C$. If hydroformylation of the olefin substrate is allowed to proceed to completion, the hydrocarbonyl is found with the reaction product. This observation by Orchin *et al.* [91] supports the postulation of the hydrocarbonyl $HCo(CO)_4$ as the active catalyst in the oxo reaction. At ordinary pressures, $HCo(CO)_4$ is very unstable [92] and decomposes rapidly into $Co_2(CO)_8$ and hydrogen. The hydrocarbonyl is relatively stable in aqueous solution, in which it is completely dissociated and behaves as a strong acid. The direct participation of hydrocarbonyl in the oxo reaction has been verified by its reaction [18,93] with olefins at room temperature under a nitrogen or hydrogen atmosphere, producing aldehydes in good yields.

Rhodium carbonyls may be substituted for cobalt carbonyls as catalysts in the oxo reaction. The use of rhodium carbonyls favors a higher proportion of straight chain-isomers at comparable temperatures. Young *et al.* [77] and Osborn *et al.* [78] have used tris(triphenylphosphine)chlororhodium(I) as a catalyst for the oxo reactions with olefins [78] and alkynes [77]. The hydroformylation of 1-hexene proceeds smoothly at 55°C and 90 atm of synthesis gas in the presence of Rh(I) catalyst at a concentration of about 1% of the substrate. The complexes 1,2,6-tris(pyridine)–Rh(III) chloride and $Rh_2Cl_2(SnCl_2 \cdot C_2H_5OH)_4$ are effective catalysts [78] for the oxo reaction of alkenes in ethanol solutions.

Iron pentacarbonyl, $Fe(CO)_5$, functions as a catalyst at temperatures above 180°C, and is accordingly used in the reduction of aldehydes to alcohols under hydroformylation conditions. Nickel tetracarbonyl, $Ni(CO)_4$, is unreactive and therefore unsuitable as an oxo catalyst.

Catalyst concentration plays an important part in determining the rate of the oxo reaction and the distribution of isomers obtained. Increasing the catalyst concentration to more than 1.6 wt% resulted [88] in a decrease in the yield of straight chain isomers. The rate of the reaction in the range of catalyst concentration of 0.05 to 0.5 wt% is a linear function [88] of the catalyst concentration.

When the reaction conditions, temperature, pressure, and catalyst concentration are held constant, the rates of hydroformylation of olefins depend on the concentration of substrate. The relative rates with various olefins decrease in the order [16], straight-chain terminal olefins > straight-chain internal olefins $\cong$ cyclic olefins > branched terminal olefins > branched internal olefins. The order reflects the steric hindrance offered by branched-chain and internal olefins for the addition of hydrogen and formyl groups across the double bond. According to the rule originally formulated by Kueleman *et al.* [94] and modified by Wender and Sternberg [95], tertiary carbon atoms participate in the oxo reaction only to a limited extent. Tertiary carbon centers that are resistant to hydroformylation may still be susceptible to homogeneous catalytic hydrogenation [96].

Regardless of the position of the double bond, a particular type of olefin generally gives the same type of products, the relative amount of each isomer in the product depending on the temperature. At 170°C, 1-heptene and 2-heptene gave equal amounts (50%) of *n*-octanal and a mixture of 2-methyl-1-heptanal and 2-ethyl-1-hexanal. At 95°C, however, the amounts of straight-chain isomers differ for the two olefins. 1-Heptene gave 72% *n*-octanal and 28% of a mixture of 2-methyl-1-heptanal and 2-ethyl-1-hexanal. 2-Heptene under the same conditions yielded 49% *n*-octanal and 51% of the branched isomers. The difference in the rates of oxo reaction of the end and internal olefins depends on the rate of complex formation of the substrate with the cobalt carbonyl catalyst.

B. PRODUCTS OF THE OXO REACTION

The primary product of the oxo reaction of a C_n olefin is a C_{n+1} aldehyde. When the oxo reaction is conducted below 120°C at 200 atm of synthesis gas, aldehydes are the main products. Conjugated dienes [97], however, yield only saturated monoaldehydes. The oxo reaction in such cases appears to take place by the reduction of one double bond and the hydroformylation of the other. Thus the oxo reaction [97] of 2,3-dimethyl-1,3-butadiene, **98**, at 145°–175°C at a pressure of 5000 lb/inch² of synthesis gas gave 3,4-dimethyl-pentanal, **101**, in 43% yield. The same product is obtained [97] by the hydro-formylation of either 2,3-dimethyl-1-butene, **100**, or 2,3-dimethyl-2-butene, **99** [Eq. (59)]. The double bond in **99** is difficult to hydroformylate directly because of the steric influence of the methyl groups [94]. Consequently the double bond migrates to the 1-position to yield **100**, which then hydro-formylates readily. Hydroformylation of butadiene **102** [Eq. (60)] gave *n*-pentanal, **103**, and 2-methylbutanal, **104**. The olefins 1,2-dimethylbutadiene and 2-methylbutadiene yield significant amounts of polymers along with the expected aldehydes.

Morikawa [98] obtained dialdehydes by the separate hydroformylation of two isolated double bonds in α,ω-diolefins such as 1,4-pentadiene, **105**, and 1,5-hexadiene, **108**, with a rhodium(III) oxide catalyst in dilute alcoholic solution below 100°C. The use of a cobalt catalyst, or the use of solvents

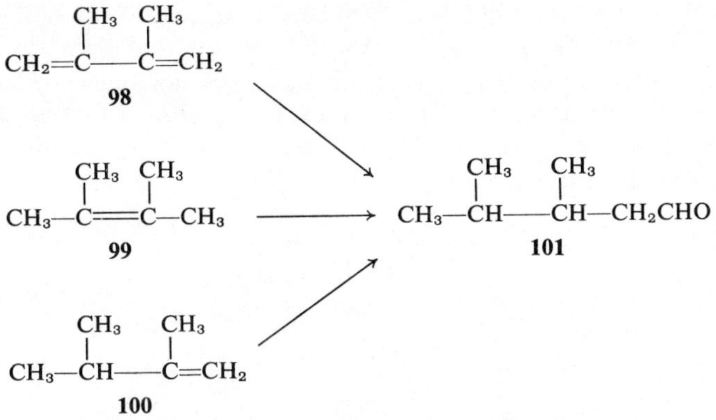

$$CH_2=C(CH_3)-C(CH_3)=CH_2$$

98

$$CH_3-C(CH_3)=C(CH_3)-CH_3$$

99

$$CH_3-CH(CH_3)-C(CH_3)=CH_2$$

100

$$CH_3-CH(CH_3)-CH(CH_3)-CH_2CHO \quad (59)$$

101

$$CH_2=CH-CH=CH_2$$

102

$$CH_3-CH_2-CH_2-CH_2-CHO$$

103

$$CH_3-CH_2-CH(CH_3)-CHO$$

104

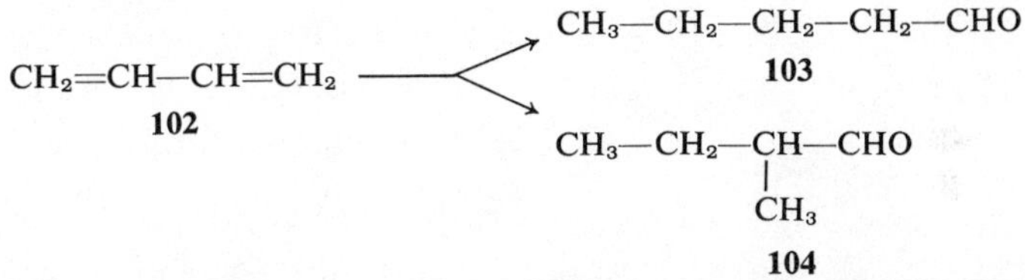

$$(60)$$

other than alcohol resulted in lower yields of aldehydes. 1,4-Pentadiene gave 1,7-heptanedial **(106)** as the major product, along with monoaldehydes and side products, in the presence of a rhodium(III) oxide catalyst. On the other hand, rhodium trichloride gave cyclohexane hydroxyaldehyde **(107)** by an intramolecular aldol condensation of the dialdehyde. The olefin 1,5-hexadiene, **108**, gave 2-methylheptanedial, **109**, as the main product of the oxo reaction [Eq. (61)]. Formation of dialdehydes from α,ω-diolefins suggests that

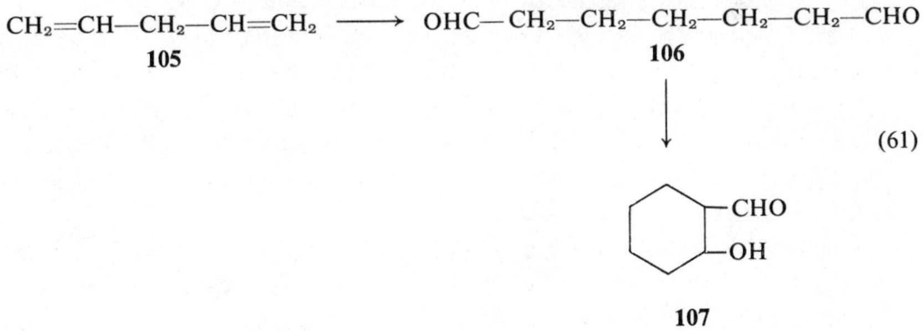

$$CH_2=CH-CH_2-CH=CH_2 \longrightarrow OHC-CH_2-CH_2-CH_2-CH_2-CH_2-CHO$$

105   **106**

$$(61)$$

107

$$CH_2=CH-CH_2-CH_2-CH_2=CH_2 \longrightarrow$$

108

$$CH_3-CH(CHO)-CH_2-CH_2-CH_2-CH_2-CHO$$

109

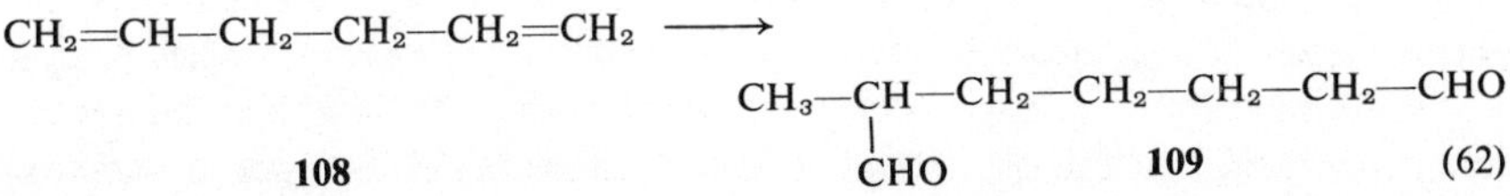

$$(62)$$

no migration of the double bonds takes place during the oxo reaction in these instances. This observation may be due to the formation of a diolefin chelate complex with the rhodium(III) chloride catalyst with subsequent 1,6-insertion of carbon monoxide as depicted by formula **111** of reaction (63). The carbonyl insertion to form a diacylrhodium(I) complex, **112**, may take place either in one step or in two separate steps.

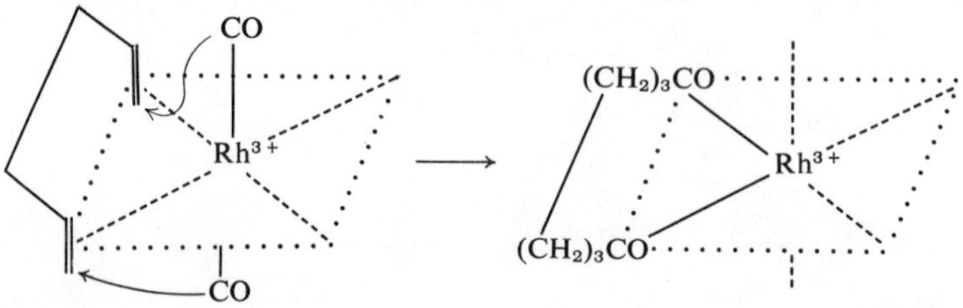

$$\text{OHC(CH}_2\text{)}_6\text{CHO} + \text{Rh(I)}$$
110

$$(63)$$

Uchida and Matsuda [99,100] were able to prepare C_{2n+1} dimeric aldehydes from C_n olefin by the addition of magnesium methylate to the reaction mixture. In the hydroformylation of propylene the highest yield (33%) of dimeric aldehyde $C_6H_{13}CHO$ was obtained at a Mg/Co molar ratio of 3:1 at 150°C under the usual synthesis gas pressure. The activation energy [100] of the oxo reaction of propylene is lowered from 22 kcal/mole to 7.5 kcal/mole at this Mg/Co ratio.

The addition of magnesium methylate to the reaction mixture causes the formation of magnesium cobalt carbonylate, $Mg[Co(CO)_4]_2$, from cobalt hydrotetracarbonyl. Thus at a given instant, both $HCo(CO)_4$ and $Mg[Co(CO)_4]_2$ are present in the solution, the former catalyzing the formation of mono C_{n+1} aldehydes and the latter catalyzing the formation of dimeric C_{2n+2} dialdehydes. The mechanism of hydroformylation of olefins in the presence of magnesium cobalt carbonylate is not fully understood and requires further detailed studies.

Hydroformylation of olefin oxides results in the formation of hydroxy aldehydes. Propylene oxide [101], **113**, undergoes hydroformylation between 80° and 100°C in the presence of cobalt octacarbonyl to give β-hydroxy-*n*-butyraldehyde, **114**, as the major product [Eq. (64)]. The temperature range of this reaction is critical. Below 80°C no appreciable reaction takes place, whereas above 100°C propylene oxide isomerizes to acetone which undergoes subsequent side reactions. Ethylene oxide probably undergoes a similar reaction, but the initial product of the reaction, β-hydroxy-n-propionaldehyde, is converted to acrolein with subsequent extensive polymerization to resinous material.

$$\underset{\textbf{113}}{\overset{\displaystyle H_2C}{\underset{\displaystyle CH_3}{\overset{|}{\underset{|}{HC}}}}\!\!>\!\!O} \quad \xrightarrow[\text{Co}_2(\text{CO})_8]{80°-100°C} \quad \underset{\textbf{114}}{CH_3\!-\!\overset{\displaystyle OH}{\overset{|}{CH}}\!-\!CH_2\!-\!CHO} \tag{64}$$

Diethyl ketone is the major product of the oxo reaction when ethylene is the substrate [102]. Heck [103,104] suggested reactions (65)–(67) as the mechanism for the formation of diethyl ketone from ethylene, carbon monoxide, and hydrogen with cobalt hydrocarbonyl as catalyst. Addition of

$$HCo(CO)_4 + CH_2\!\!=\!\!CH_2 \quad \rightleftharpoons \quad \underset{\textbf{115}}{CH_3CH_2Co(CO)_4} \tag{65}$$

$$CH_3CH_2Co(CO)_4 + CO + CH_2\!\!=\!\!CH_2 \quad \longrightarrow \quad \underset{\textbf{116}}{CH_3CH_2COCH_2CH_2Co(CO)_4} \tag{66}$$

$$CH_3CH_2COCH_2CH_2Co(CO)_4 + H_2 \quad \longrightarrow \quad CH_3CH_2COCH_2CH_3 + HCo(CO)_4 \tag{67}$$

the hydrocarbonyl to ethylene results in the formation of ethylcobalt tetra-carbonyl, **115**. Olefin and carbon monoxide insertion reactions in **115** give rise to the highly reactive β-acylethylcobalt tetracarbonyl, **116**, which undergoes hydrogenolysis to 3-pentanone. It was assumed [103] that β-acylethylcobalt tetracarbonyl is so highly reactive that there is no conversion of the tetra-carbonyl to β-propionylpropionylcobalt tricarbonyl, $CH_3CH_2COCH_2CH_2$-$COCo(CO)_3$, which would give an aldehyde on hydrogenolysis.

Bertrand et al. [105] have suggested the formation of a ketone by the inter-action of an acylcobalt tricarbonyl and alkylcobalt tetracarbonyl according to reaction (68). Formation of a ketone is favored when the concentration of

$$RCH_2CH_2COCo(CO)_3 + R'Co(CO)_4 \quad \longrightarrow \quad RCH_2CH_2COR' + Co_2(CO)_7 \tag{68}$$

the hydrocarbonyl is kept at a minimum by extensive conversion to alkyl-cobalt tetracarbonyl in the reaction mixture; otherwise an aldehyde is the major product. The reaction of 1,4-pentadiene with hydridocobalt tetra-carbonyl at ambient temperature and pressure gave a mixture of 2-methyl-cyclopentanone **123**, along with small quantities of cyclohexenone **121** and 2-methylcyclopenten-3-one **122**. The reaction of a conjugated olefin, 1,3-buta-diene, with $HCo(CO)_4$ requires heating to 120°C at a carbon monoxide pressure of 1200 lb/inch², to yield 5-nonanone exclusively.

Heck [103] has explained the formation of cyclic products in the reaction of 1,4-pentadiene with hydridocobalt tetracarbonyl on the basis of the formation of a cyclic π-olefin complex from 5-hexenoylcobalt tetracarbonyl which undergoes internal olefin insertion and disproportionation to give various products, as indicated by reaction sequence (69)–(72).

$$CH_2{=}CH{-}CH_2{-}CH{=}CH_2 + HCo(CO)_4 \;\rightleftharpoons$$
$$CH_2{=}CH{-}CH_2{-}CH_2{-}CH_2{-}Co(CO)_4 \quad (69)$$

$$CH_2{=}CH{-}CH_2{-}CH_2{-}CH_2{-}Co(CO)_4 + CO \;\rightleftharpoons$$
$$CH_2{=}CH{-}CH_2{-}CH_2{-}CH_2{-}COCo(CO)_4 \quad (70)$$

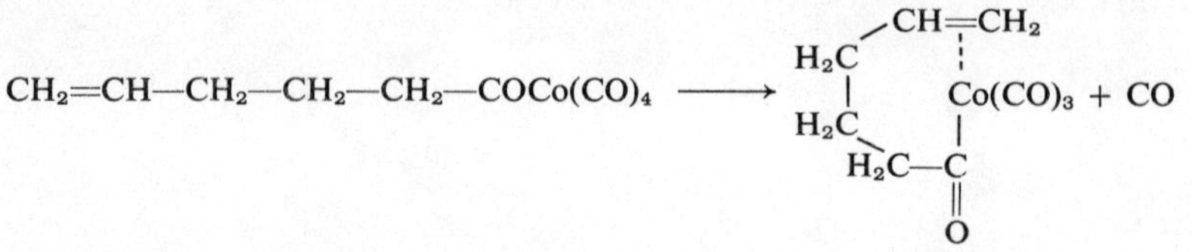

$$CH_2{=}CH{-}CH_2{-}CH_2{-}CH_2{-}COCo(CO)_4 \longrightarrow \quad Co(CO)_3 + CO \quad (71)$$

117   118

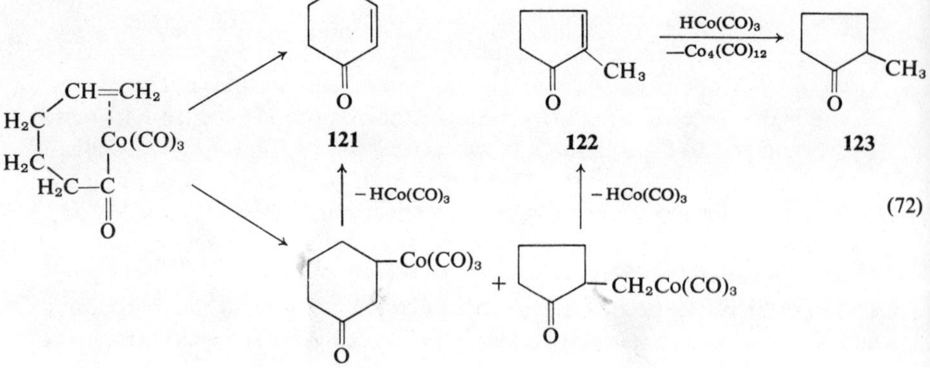

(72)

111   119   120

1. Secondary Reaction Products

A large number of secondary products can be obtained in hydroformylation through subsequent interactions of the active carbonyl groups of the aldehydes. Table II lists some of the secondary products obtained in the oxo reaction. Some of the reactions listed in Table II will be considered in detail in Section IV.

TABLE II

SECONDARY PRODUCTS OBTAINED IN THE OXO REACTION

Oxo intermediate	Reaction	Catalyst	Conditions	Secondary products	Ref.
Saturated aldehydes	Hydrogenation under oxo conditions	Co(CO)₃ or Fe(CO)₅	CO + 2 H₂ at 190–200°C 150–300 atm	Saturated alcohols	99 106
Saturated aldehydes	Hydrogenation under oxo conditions	(PPh₃)₃RhCl	CO + 2 H₂ at 50°C, 50 atm	Saturated alcohols	78
Saturated aldehydes	Reaction with HCHO	—	Special Tishchenko reaction	Formates	74
Saturated aldehydes	Reaction with alcohols	Co₂(CO)₈	CO + H₂ at 150–200 atm 160°C	Esters	74
Saturated aldehydes	Reaction with alcohols or esters	Co₂(CO)₈	CO + H₂ at 150–200 atm 160°C	Acetals (prevents side reactions of aldehydes)	74
Saturated aldehydes	Reaction with amines	Co₂(CO)₈	CO + H₂ at 150–200 atm 160°C	Amides, nitriles	85
Saturated aldehydes	Hydroamination	Co₂(CO)₈	—	Amines	85
Saturated aldehydes	Aldolization	Co₂(CO)₈	Oxo conditions	Aldols and higher alcohols	74
Unsaturated aldehydes	Hydrogenation	Co₂(CO)₈	Oxo conditions	Hydroxy aldehydes	85
Aromatic aldehydes and alcohols	Hydrogenation	Co₂(CO)₈	Oxo conditions	Saturated hydrocarbons	108

C. KINETICS AND MECHANISM OF THE OXO REACTION

From the values of heats and entropies of formation of ethylene, carbon monoxide, and hydrogen at 25°C, $\Delta H°$ and $\Delta G°$ for reaction (73) were calculated [85] as -34.8 and -17.47 kcal/mole, respectively. Natta *et al.* [109]

$$CH_2{=}CH_2 + CO + H_2 \longrightarrow CH_3CH_2CHO \tag{73}$$

reported an experimental value for $\Delta G°$ of this reaction of -14.46 kcal/mole at 25°C. Since the standard free energy of hydrogenation of ethylene [110] at 25°C is -22.61 kcal/mole, it is seen that at 1 atm and 25°C, and under equilibrium conditions, hydrogenation of olefins would be favored over hydroformylation. However, under oxo conditions the reaction of hydrogen and carbon monoxide with olefins gives exclusively hydroformylation products.

The kinetics of hydroformylation was originally reported [111] as heterogeneous, but later experiments [112] indicated that the hydroformylation reaction is truly homogeneous; the rate of the reaction increases with time as cobalt goes into solution in the form of cobalt carbonyl. Eventually cobalt carbonyls were directly used [113,114] at 110°C and at 100–380 atm of synthesis gas in the rate studies. Under these conditions, the rate of the oxo reaction was found to be first order with respect to olefin [112] (e.g., cyclohexene) and to the catalyst, and was independent of the total pressure of synthesis gas in the 100–380 atm range.

In the oxo reaction of diisobutylene [115] at 150°C, the rate of the reaction was found to be always proportional to the olefin concentration and to the cobalt carbonyl catalyst concentration over a range of partial pressures of hydrogen and carbon monoxide from 1 atm to 3 atm. At constant carbon monoxide pressure, the rate of the oxo reaction increased rapidly with increasing hydrogen pressure. At constant partial pressure of hydrogen, the rate increases with increasing partial pressure of carbon monoxide up to about 10 atm of carbon monoxide but decreases at pressures more than 10 atm. This effect of the variation of partial pressure of carbon monoxide was studied [113] in the oxo reaction of cyclohexene in toluene solution at $110° \pm 1°C$, with dicobalt octacarbonyl as a catalyst. The following empirical relationship between initial rate of the oxo reaction [Eq. (73)] and partial pressures of hydrogen and carbon monoxide, proposed by Martin [115], indicate inverse dependence of rate on partial pressure of carbon monoxide, as represented by Eq. (74).

$$\text{Initial rate} = \frac{275 p_{H_2}}{0.22 p_{H_2} + p_{CO}} \tag{74}$$

The work of Iwanaga [116,117] has further elucidated the effects of partial pressure of carbon monoxide and hydrogen on the initial rate of the formation of cobalt hydrocarbonyl, $HCo(CO)_4$ [116], the active form of the catalyst, and on the rate of the oxo reaction [117]. The initial rate of formation of cobalt hydrocarbonyl increases with increasing partial pressure of hydrogen and decreases with increasing pressure of carbon monoxide. The effects of partial pressure of carbon monoxide and hydrogen on the initial rate of formation of $HCo(CO)_4$ thus parallel their effects on the rate of the oxo reaction reported by Natta et al. [113]. At a particular time during the oxo reaction, as determined by a titration procedure [118], only traces of cobalt hydrocarbonyl were present in the reaction of cyclohexene at 120°C in toluene solution. The retarding effect of increasing partial pressure of carbon monoxide on the rate of the oxo reaction was explained by Metlin et al. [119] on the basis of probable formation of catalytically inactive dicobalt nonacarbonyl, $Co_2(CO)_9$, from dicobalt octacarbonyl and carbon monoxide according to Eq. (75). The formation of $Co_2(CO)_9$ under oxo conditions has,

$$Co_2(CO)_8 + CO \rightleftharpoons Co_2(CO)_9 \qquad (75)$$

however, been disputed by Iwanaga [116]. The equilibrium concentration of $HCo(CO)_4$ in the reaction mixture of dicobalt octacarbonyl and hydrogen was measured under varying partial pressures of carbon monoxide at a constant partial pressure of hydrogen. Depending on the partial pressure of hydrogen employed, the equilibrium concentration of the cobalt hydrocarbonyl is reached within 40–75 minutes at 120°C. It was further reported that the equilibrium concentration of $HCo(CO)_4$ was not affected by the partial pressure of carbon monoxide, the latter influencing only the initial rates of formation of $HCo(CO)_4$. If any $Co_2(CO)_9$ is formed in accordance with Eq. (75), the reduction in $Co_2(CO)_8$ concentration with increasing partial pressure of carbon monoxide should reduce the equilibrium concentration of $HCo(CO)_4$, as indicated by Eq. (76). Since no such variation was noticed, it was concluded that no $Co_2(CO)_9$ is formed during the oxo reaction.

Mechanisms suggested prior to 1956 by Wender et al. [16,95] and by Martin [115] assigned no part to cobalt hydrocarbonyl, which has been demonstrated to be present under oxo conditions by Orchin et al. [91] and by IR evidence by Marko [120]. Natta et al. [114] also suggested the possibility of the active participation of cobalt hydrocarbonyl in the oxo reaction. The formation and rapid utilization of $HCo(CO)_4$ in the oxo reaction led to the conclusion that the hydrocarbonyl must be involved through the formation of an olefin–hydrocarbonyl complex [121]. The reaction sequence proposed by Kirch and Orchin [121] is given in Eqs. (76)–(78). The reactions suggested by Kirch and Orchin [121] were modified by Karapinka and Orchin [18] to Eqs. (76), (79), and (80). Breslow and Heck [122] and Heck [103] have postulated the mechanism given in Eqs. (76) and (81)–(84) for the oxo reaction. Inhibition of the oxo reaction by carbon monoxide was explained

$$Co_2(CO)_8 + H_2 \rightleftharpoons 2 HCo(CO)_4 \qquad (76)$$
$$\textbf{124} \qquad\qquad \textbf{125}$$

$$2 HCo(CO)_4 + RCH{=}CH_2 + CO \longrightarrow \{[HCo(CO)_4]_2 \cdot RCH{=}CH_2 \cdot CO\} \quad (77)$$

$$\{[HCo(CO)_4]_2 \cdot RCH{=}CH_2 \cdot CO\} \longrightarrow Co_2(CO)_8 + RCH_2CH_2CHO \qquad (78)$$

$$Co_2(CO)_8 + H_2 \rightleftharpoons 2 HCo(CO)_4 \qquad (76)$$

$$HCo(CO)_4 + CO + RCH{=}CH_2 \longrightarrow RCH_2CH_2COCo(CO)_4 \qquad (79)$$

$$HCo(CO)_4 + RCH_2CH_2COCo(CO)_4 \longrightarrow RCH_2CH_2CHO + Co_2(CO)_8 \qquad (80)$$

[122] on the basis of the equilibrium between $HCo(CO)_4$ and the more reactive hydridocobalt tricarbonyl, $HCo(CO)_3$. The hydro tricarbonyl complex was considered to add to the olefin in the next step to form alkylcobalt tricarbonyl. Carbon monoxide insertion in the alkylcobalt tricarbonyl forms an acylcobalt tricarbonyl in equilibrium with alkylcobalt tetracarbonyl. It was assumed that in the last step the coordinatively unsatured acylcobalt

$$Co_2(CO)_8 + H_2 \rightleftharpoons 2 HCo(CO)_4 \qquad (76)$$
$$\textbf{124} \qquad\qquad \textbf{125}$$

$$HCo(CO)_4 \rightleftharpoons HCo(CO)_3 + CO \qquad (81)$$
$$\textbf{126}$$

$$HCo(CO)_3 + RCH{=}CH_2 \longrightarrow RCH_2CH_2Co(CO)_3 \qquad (82)$$

$$RCH_2CH_2Co(CO)_3 + CO \longrightarrow RCH_2CH_2COCo(CO)_3 \qquad (83)$$

$$RCH_2CH_2COCo(CO)_3 + H_2 \longrightarrow RCH_2CH_2CHO + HCo(CO)_3 \qquad (84)$$

tricarbonyl (in preference to alkylcobalt tetracarbonyl) reacts with molecular hydrogen to form the products and regenerate the catalyst, $HCo(CO)_3$. It should be noted that the formation of $HCo(CO)_3$ as an intermediate was questioned by Bertrand et al. [105] on the basis of mass spectroscopic analyses in which no hydrogen-containing species related to the tricarbonyl fragment were detected.

The last steps in the mechanisms of Karapinka and Orchin [18] and of Breslow and Heck [122] involve cobalt hydrotetracarbonyl and molecular hydrogen, respectively. There is no unequivocal evidence yet in favor of the participation of a single hydrogenation species, $HCo(CO)_4$ or hydrogen, in oxo product formation. Strong evidence is also lacking for the presence of coordinatively unsaturated species as participants in the oxo reaction.

Heck [103] has suggested that acylcobalt tricarbonyl, **127**, adds molecular hydrogen to form the intermediate tricarbonyl acyldihydridocobalt, **128** which decomposes to the aldehyde and $HCo(CO)_3$ by a hydride shift to the coordinated acyl group.

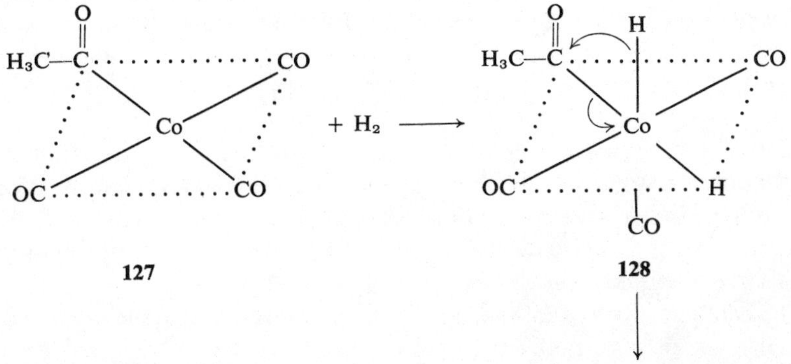

$$CH_3CHO + HCo(CO)_3$$

(85)

The conversion of dicobalt octacarbonyl to cobalt hydrotetracarbonyl was fully substantiated by the infrared evidence of Marko [120] and direct isolation of $HCo(CO)_4$ in the absence of the substrate by Orchin et al. [91] and by Iwanaga [116].

Evidence pertaining to the dissociation of dicobalt octacarbonyl is provided by the catalytic effect of pyridine and pyridine homologs on the rate of the oxo reaction when added in small quantities. In the presence of a molar quantity of pyridine equal to that of the catalyst, the rate of the oxo reaction is doubled [123,124]. The more basis aliphatic tertiary amines, such as triethylamine and tributylamine, however, cause a reduction of the rate. The catalytic effect of small quantities of pyridine was explained by Iwanaga [123] on the basis of a pre-equilibrium step involving formation of a $Co_2(CO)_8-$ base complex and subsequent reaction of this complex with hydrogen to form $HCo(CO)_4$, as indicated in reactions (86) and (87). Large quantities of

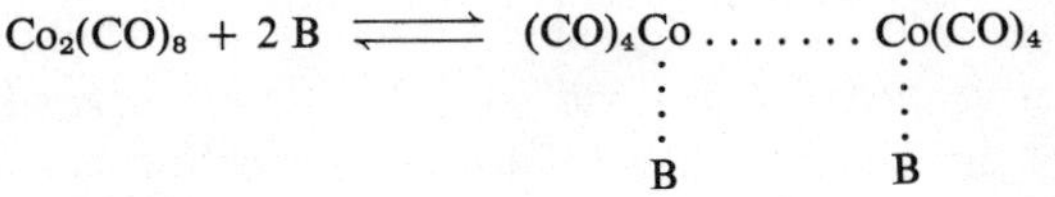

$$Co_2(CO)_8 + 2\ B\ \rightleftharpoons\ (CO)_4Co \cdots\cdots Co(CO)_4 \qquad (86)$$

$$\underset{124}{}\qquad\qquad \underset{\substack{B \\ 129}}{} \qquad \underset{B}{}$$

$$(CO)_4Co \cdots\cdots Co(CO)_4 + H_2\ \longrightarrow\ 2\ HCo(CO)_4 + 2\ B \qquad (87)$$
$$\underset{B}{}\qquad\qquad \underset{B}{}$$

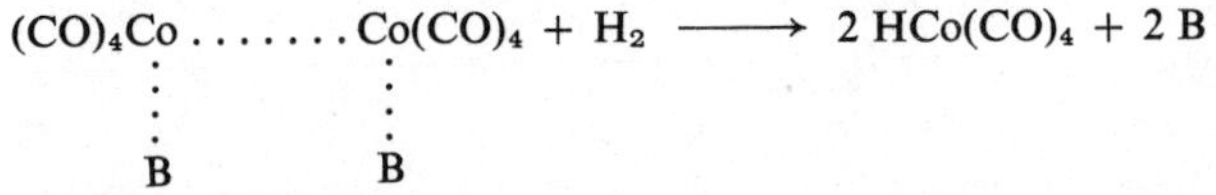

pyridine inhibit the reaction by removing the $HCo(CO)_4$ in the form of its pyridinium salt [15]. Aliphatic amines, which are more basic than pyridine,

$$HCo(CO)_4 + B\ \rightleftharpoons\ [BH^+][Co(CO)_4]^- \qquad (88)$$

form stable salts of the type indicated in reaction (88) and thus lower the degree of formation of the reactive hydrocarbonyl.

The activation effect of protic solvents [125] on the rate of the oxo reaction of methyl acrylate further supports the concept of formation of a complex with added base of the type indicated by **129**. The rate of the oxo reaction varies in the order alcohols > acetone > toluene with these substances as solvents. Acceleration of the rate of hydroformylation of olefin oxides by ethyl and butyl alcohols has been ascribed [126] to preequilibrium complex formation with the solvent, and consequent polarization of the Co—Co bond. Highly polar solvents such as ethylene glycol [Eq. (89)] and formamide inhibit the oxo reaction completely. In the case of ethylene glycol, a complex salt is formed which is stabilized in the highly polar solvent, and consequently no further reaction takes place.

The catalytic effect of thiophenol on the oxo reaction may be explained in the same manner as the influence of alcohols. Morikawa [98] reported that the activation energy of hydroformylation of cyclohexene is reduced from 30 to 13 kcal/mole by the addition of thiophenol to the reaction mixture. Also, the induction period of the reaction is reduced from 85 to 22 minutes by the presence of thiophenol.

$$2\ Co_2(CO)_8 + 3\ C_2H_6O_2 \longrightarrow [Co(C_2H_6O_2)_3][Co(CO)_4]_3 + 4\ CO \qquad (89)$$

Additional support for the participation of cobalt hydrocarbonyl in the oxo reaction is provided by the inhibition [127] of the reaction in the presence of alkynes that are known to react with $Co_2(CO)_8$ to produce inactive complexes of the type (alkyne)$Co_2(CO)_6$.

The formation of hydridocobalt tricarbonyl is based on kinetic evidence for the inhibition of the oxo reaction by carbon monoxide. There is [18,122] no direct evidence, however, for the formation of the coordinately unsaturated species, $HCo(CO)_3$, in solution. According to Pino *et al.* [128] both $HCo(CO)_4$ and $HCo(CO)_3$ may take part in a mechanism of the type suggested by Breslow and Heck [122], and they suggested that the relative concentrations of branched-chain and straight-chain isomers formed depend on the equilibrium concentration of the alkylcobalt tetracarbonyl and acylcobalt tricarbonyl, at a particular temperature and partial pressure of carbon monoxide. Pino *et al.* [128] studied the effect of increasing partial pressure of carbon monoxide on the ratio of branched- to straight-chain isomers produced in the hydroformylation of propylene at constant temperature and constant hydrogen pressure. To avoid complications due to possible side reactions of the carbonyl compound, the carbon monoxide was buffered with ethyl orthoformate. The amount of straight-chain isomer increased with partial pressure of carbon monoxide up to a limiting value characteristic of the temperature employed. Thus at 180° and 80°C, the limiting percent yields of the straight-chain isomer were, respectively, 73.3% and 67.7%. The partial pressures of carbon monoxide corresponding to the limiting yields of the straight-chain isomer were 147 and 80 atm, respectively. Further increase of carbon monoxide pressure produced little change in the limiting yield of the straight-chain isomer. The variation of the ratio of straight- to branched-chain isomers was rationalized on the basis of the equilibria of Eqs. (91)–(93). At

$$CH_3CH{=}CH_2 + HCo(CO)_4 \longrightarrow C_3H_7Co(CO)_4 \qquad (90)$$

$$C_3H_7Co(CO)_4 \;\rightleftharpoons\; C_3H_7Co(CO)_3 + CO \qquad (91)$$

$$\begin{array}{c} H_3C \\ \diagdown \\ \diagup \\ H_3C \end{array}\!\!CHCo(CO)_4 \;\rightleftharpoons\; CH_3CH_2CH_2Co(CO)_4 \qquad (92)$$

$$\begin{array}{c} H_3C \\ \diagdown \\ \diagup \\ H_3C \end{array}\!\!CHCo(CO)_4 \;\rightleftharpoons\; CH_3CH_2CH_2Co(CO)_3 + CO \qquad (93)$$

constant temperature, an increase in partial pressure of carbon monoxide at or above a minimum value influences the equilibrium between $C_3H_7Co(CO)_3$ and $C_3H_7Co(CO)_4$, Eq. (91), by increasing the concentration of the tetra-carbonyl. Although the effect of pressure on the equilibria between straight-chain and branched-chain tri- and tetracarbonyls is not fully understood, it is nevertheless assumed on steric grounds that increase of pressure favors the formation of straight-chain tetracarbonyl. Thus, above a limiting pressure of carbon monoxide (greater than 40 atm), the propylcobalt tetracarbonyl species at equilibrium is believed to contain more n-propyl isomer than does the propylcobalt tricarbonyl species. This results in propylcobalt tetra-carbonyl producing a higher proportion of the straight-chain aldehyde in the final product.

These studies of the effect of partial pressure of carbon monoxide on the ratio of straight- to branched-chain isomers were extended by Piacenti *et al.* [129] to other alkene substrates, including, 1-butene, *cis*-2-butene, 1-pentene, 4-methyl-1-pentene, and 2-pentenes. At low carbon monoxide pressure of 1–2 atm, the relative amounts of straight- to branched-chain hydroformyla-tion products obtained from isomeric terminal and internal olefins are about the same. The ratio of straight- to branched-chain products increases with carbon monoxide pressure for both the terminal and internal olefins at temperatures of 100°–116°C.

The isomeric composition of the hydroformylation products also depends on the solvent [130]. Although the effect of the solvent on the ratio of straight- to branched-chain isomers is not well characterized, it may be due partly to solvation effects and partly to differences in the solubility of carbon monoxide in different solvent systems.

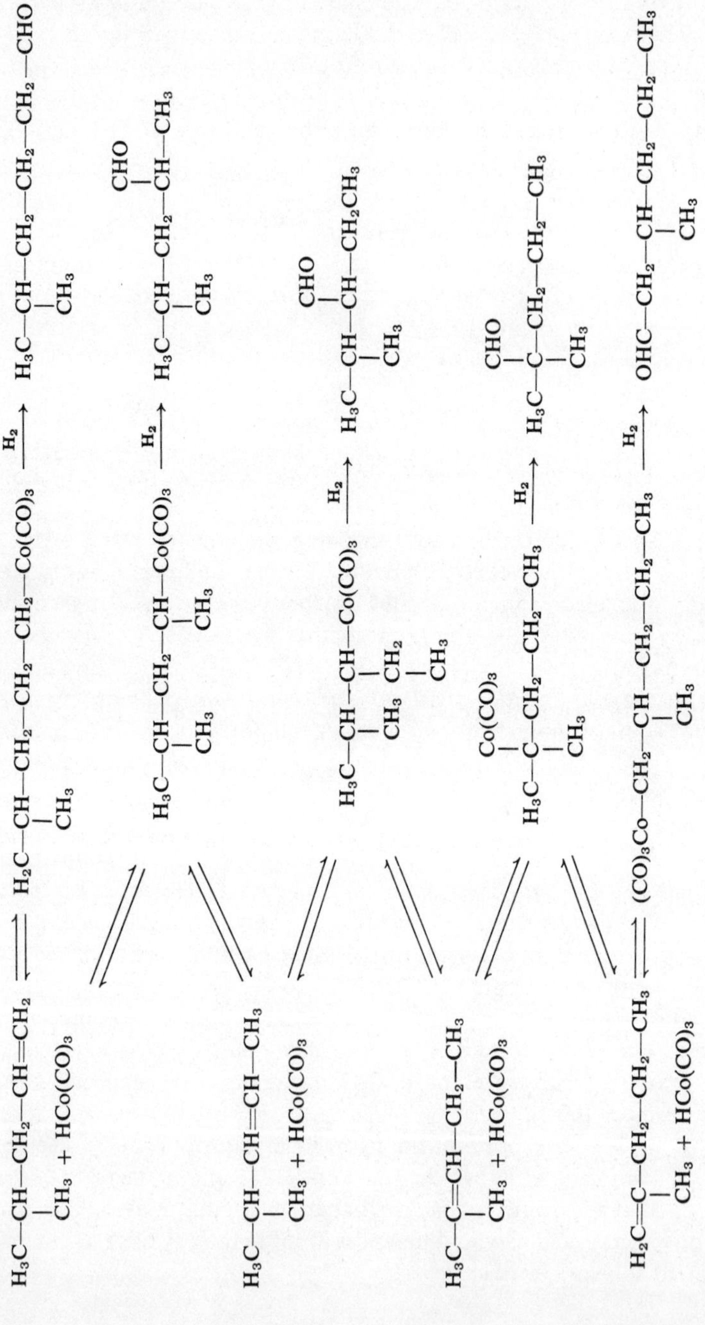

$$H_3C-CH-CH_2-CH=CH_2 \ (+ \ HCo(CO)_3)$$
$$\quad\quad\quad |$$
$$\quad\quad\quad CH_3$$

$$\rightleftharpoons \ H_2C-CH-CH_2-CH_2-CH_2-Co(CO)_3 \ \xrightarrow{H_2} \ H_3C-CH-CH_2-CH_2-CH_2-CHO \quad (94a)$$
$$\quad\quad\quad\quad |\quad\quad\quad\quad\quad\quad\quad\quad\quad\quad\quad\quad\quad\quad |$$
$$\quad\quad\quad\quad CH_3\quad\quad\quad\quad\quad\quad\quad\quad\quad\quad\quad\quad\quad CH_3$$

$$H_3C-CH-CH=CH-CH_3 \ (+ \ HCo(CO)_3)$$
$$\quad\quad |$$
$$\quad\quad CH_3$$

$$\rightleftharpoons \ H_3C-CH-CH_2-CH-Co(CO)_3 \ \xrightarrow{H_2} \ H_3C-CH-CH_2-CH-CH_3 \quad (94b)$$
$$\quad\quad\quad\quad |\quad\quad\quad\quad\quad |\quad\quad\quad\quad\quad\quad\quad |\quad\quad\quad\quad |$$
$$\quad\quad\quad\quad CH_3\quad\quad\quad CH_3\quad\quad\quad\quad\quad CH_3\quad\quad\quad CHO$$

$$H_3C-C=CH-CH_2-CH_3 \ (+ \ HCo(CO)_3)$$
$$\quad\quad |$$
$$\quad\quad CH_3$$

$$\rightleftharpoons \ H_3C-CH-CH-Co(CO)_3 \ \xrightarrow{H_2} \ H_3C-CH-CH-CH_2CH_3 \quad (94c)$$
$$\quad\quad\quad\quad |\quad\quad |\quad\quad\quad\quad\quad\quad\quad\quad |\quad\quad |$$
$$\quad\quad\quad\quad CH_3\ CH_2\quad\quad\quad\quad\quad\quad CH_3\ CHO$$
$$\quad\quad\quad\quad\quad\quad\quad |$$
$$\quad\quad\quad\quad\quad\quad\quad CH_3$$

$$H_3C-C=CH-CH_2-CH_3 \ (+ \ HCo(CO)_3)$$
$$\quad\quad |$$
$$\quad\quad CH_3$$

$$\rightleftharpoons \ H_3C-C-CH_2-CH_2-CH_3 \ \xrightarrow{H_2} \ H_3C-C-CH_2-CH_2-CH_3 \quad (94d)$$
$$\quad\quad\quad\quad |\quad\quad\quad\quad\quad\quad\quad\quad\quad\quad\quad\quad |$$
$$\quad\quad\quad Co(CO)_3\ CH_3\quad\quad\quad\quad\quad CHO\ CH_3$$

$$H_2C=C-CH_2-CH_2-CH_3 \ (+ \ HCo(CO)_3)$$
$$\quad\quad\quad |$$
$$\quad\quad\quad CH_3$$

$$\rightleftharpoons \ (CO)_3Co-CH_2-CH-CH_2-CH_2-CH_3 \ \xrightarrow{H_2} \ OHC-CH_2-CH-CH_2-CH_2-CH_3 \quad (94e)$$
$$\quad\quad\quad\quad\quad\quad\quad\quad\quad |\quad\quad\quad\quad\quad\quad\quad\quad\quad\quad\quad\quad\quad\quad\quad |$$
$$\quad\quad\quad\quad\quad\quad\quad\quad\quad CH_3\quad\quad\quad\quad\quad\quad\quad\quad\quad\quad\quad\quad CH_3$$

Brewis [131] has studied the hydroformylation of propylene at 100°–250°C and from 250 to 2500 atm of carbon monoxide. Below 130°C, total pressure has no influence on product distribution but in the temperature range of 130°–200°C, total pressure has a marked effect on the ratio of straight- to branched-chain isomers. At 160°C, the yields of the straight-chain isomer n-butyraldehyde at 250 and 750 atm of carbon monoxide are, respectively, 56.5% and 76.0%. The yield of n-butyraldehyde reaches a maximum of 76.7% at 1000 atm and then drops to 67.2% at 2500 atm. The decrease in the yield at very high pressures has been explained on the basis of a compressibility factor which shifts the equilibrium between n-propylcobalt tricarbonyl and isopropylcobalt tricarbonyl towards the latter isomer on the assumption that the molar volume of the isopropyl group is less than that of the n-propyl group.

The isomerization of olefins in the oxo reaction depends on the nature of the olefin and the temperature. Metal-catalyzed olefin isomerization has been considered in detail in Chapter 2. Both the end olefins and internal olefins give the same final products, and the relative rates of the reactions seem to depend on the relative rates of the olefin insertion step. Once an alkyl-cobalt tricarbonyl or tetracarbonyl is formed, isomerization requires very little energy and the reaction then proceeds rapidly. Thus, various isomeric 4-methylpentenes [132] give the same end products: 5-methylhexanal, 2,4-dimethylpentanal, 2-ethyl-3-methylbutanal, 2,2-dimethylpentanal, and 3-methylhexanal, as indicated, by the reaction sequence (94a)–(94e). Similarly, 1-, 2-, and 3-hexene form mixtures of the same isomeric aldehydes, n-heptanal, 2-methylpentanal, and 2-ethylbutanal [132].

Yagupsky et al. [79] and Brown and Wilkinson [80] have studied the kinetics and mechanism of the hydroformylation of alkenes with hydrido-carbonyltris(triphenylphosphine)rhodium, $RhH(CO)(PPh_3)_3$, 130. From 1-alkenes high ratios of straight-chain to branched-chain (95% straight chain) aldehydes were produced [80]. The mechanism proposed [80] is illustrated by Eqs. (95)–(101). Reaction of 130 with a 1:1 mixture of carbon monoxide and hydrogen at ambient room temperature and pressure results [79] in the formation of the catalytically active species 131. Reaction with the alkene in step (96) form the complex 132 which by a hydride transfer in reaction (97) forms the metal–alkyl complex 133. Carbonyl insertion in the metal–alkyl bond of 133 gives rise to the acyl complex 134 in step (98). The complex 134 forms the dihydride 135 in the rate-determining step (99) ($k < k_{-1}$). Complex 135 then gives the aldehyde and 131. Reaction (99) is partly inhibited by carbon monoxide because of the formation of the unreactive compound 136. The rate of hydroformylation for alkenes decreases in the order 1-alkenes > 2-alkenes > branched 1-alkenes, indicating a strong steric effect of the alkene structure on the rate of hydroformylation.

$$\text{RhH(CO)(PPh}_3)_3 + \text{CO} \; \rightleftharpoons \; \text{RhH(CO)}_2(\text{PPh}_3)_2 + \text{PPh}_3 \qquad (95)$$
$$\qquad\quad \textbf{130} \qquad\qquad\qquad\qquad\quad \textbf{131}$$

$$\text{RhH(CO)}_2(\text{PPh}_3)_2 + \text{alkene} \; \rightleftharpoons \; \text{RhH(alkene)(CO)}_2(\text{PPh}_3)_2 \qquad (96)$$
$$\qquad\qquad\qquad\qquad\qquad\qquad\qquad \textbf{132}$$

$$\text{RhH(alkene)(CO)}_2(\text{PPh}_3)_2 \; \rightleftharpoons \; \text{RhR(CO)}_2(\text{PPh}_3)_2 \qquad (97)$$
$$\qquad\qquad\qquad\qquad\qquad \textbf{133}$$

$$\text{RhR(CO)}_2(\text{PPh}_3)_2 \; \rightleftharpoons \; \text{Rh(COR)(CO)(PPh}_3)_2 \qquad (98)$$
$$\qquad\qquad\qquad\qquad \textbf{134}$$

$$\text{Rh(COR)(CO)(PPh}_3)_2 + \text{H}_2 \; \underset{k_{-1}}{\overset{k}{\rightleftharpoons}} \; \text{Rh(COR)H}_2(\text{CO)(PPh}_3)_2 \qquad (99)$$
$$\qquad \textbf{134} \qquad\qquad\qquad\qquad\qquad\qquad \textbf{135}$$

$$\text{Rh(COR)H}_2(\text{CO)(PPh}_3)_2 \; \longrightarrow \; \text{RhH(CO)(PPh}_3)_2 + \text{RCHO} \qquad (100)$$
$$\qquad\qquad\qquad\qquad\qquad\qquad\qquad \textbf{131}$$

$$\text{Rh(COR)(CO)(PPh}_3)_2 + \text{CO} \; \longrightarrow \; \text{Rh(COR)(CO)}_2(\text{PPh}_3)_2 \qquad (101)$$
$$\qquad\qquad\qquad\qquad\qquad\qquad\qquad \textbf{136}$$

III. Hydroformylation of Alkynes

Most of the published work on the oxo reaction has been centered on the reactions of alkenes, and relatively little attention has been given to the corresponding reactions of alkynes. Greenfield et al. [127] carried out the oxo reaction on 1-pentyne at 130°C by the use of $Co_2(CO)_8$ as the catalyst. Isolation of the products was carried out by converting them to the corresponding alcohols, since the alcohols are less reactive than aldehydes and thus easier to separate. A 6% yield of 1-hexanal and a 5.5% yield of 2-methyl-1-pentanal were obtained. The remainder of the material was a high-boiling mixture of C_{12} esters. The corresponding alkene, 1-pentene, reacts at temperatures as low as 90°C under similar conditions. It is therefore concluded that the alkynes react more slowly under oxo conditions than do the corresponding alkenes. The C=C double bond is generally more reactive than a comparably situated triple bond towards electrophilic reagents and free radicals. The reverse behavior is observed for nucleophilic reagents.

Diphenylacetylene reacts at about 170°C under oxo conditions to give 1,2-diphenylethane in 80% yield, as indicated in reaction sequence (102).

$$C_6H_5-C{\equiv}C-C_6H_5 \; \longrightarrow \; C_6H_5-CH{=}CH-C_6H_5 \; \longrightarrow \; C_6H_5-CH_2CH_2-C_6H_5$$
$$(102)$$

cis-Stilbene reacts similarly to give only 1,2-diphenylethane. Aromatic alkynes are thus hydrogenated [74] with intermediate formation of alkenes, under conditions that would be expected to lead to hydroformylation.

Hydroformylation of 1-pentyne may take place by two possible pathways. In reaction sequence (103) 1-pentyne is first hydrogenated to 1-pentene [in a manner analogous to reaction (102)], which is then hydroformylated to hexanal and 2-methylpentanal. In an alternative sequence, Eq. (104), 1-pentyne is first hydroformylated to an α,β-unsaturated aldehyde, and the double bond and carbonyl group are hydrogenated in subsequent steps.

$$CH_3-CH_2-CH_2-C\equiv CH \longrightarrow$$

$$CH_3-CH_2-CH_2-CH=CH_2 \begin{array}{c} \nearrow CH_3CH_2CH_2\overset{\overset{\displaystyle CH_3}{\displaystyle |}}{CH}-CHO \\ \\ \searrow CH_3CH_2CH_2CH_2CH_2CHO \end{array} \qquad (103)$$

$$CH_3-CH_2-CH_2-C\equiv CH \longrightarrow$$

$$CH_3-CH_2-CH_2-CH=CH-CHO \longrightarrow CH_3CH_2CH_2CH_2CH_2CHO \qquad (104)$$

$$CH_3-CH_2-CH_2-\overset{\overset{\displaystyle CH_2}{\displaystyle \|}}{C}-CHO \longrightarrow CH_3CH_2CH_2\overset{\overset{\displaystyle CH_3}{\displaystyle |}}{CH}-CHO$$

The active catalyst for the oxo reaction of alkynes is considered to be hydridocobalt tetracarbonyl, as indicated by the direct interaction of acetylene [127] with the hydrocarbonyl to give a π complex, from which propionaldehyde is obtained. Some of the acetylene complex of dicobalt hexacarbonyl is also formed in this reaction, possibly by the interaction of acetylene directly with $Co_2(CO)_8$.

Tris(triphenylphosphine)chlororhodium(I) [77] has been used as a catalyst in the oxo reaction of alkynes. The reaction takes place under conditions milder than those used in the conventional oxo reaction with cobalt carbonyls. With this catalyst, hydroformylation of an alkyne is quantitative, and the distribution of the aldehyde isomers is similar to that obtained in the cobalt carbonyl-catalyzed reaction. The oxo reaction of alkynes with rhodium(I) catalysts probably takes place through a mechanism similar to that proposed by Breslow and Heck [122] and by Karapinka and Orchin [18] described in Section II.

IV. Hydrogenation under Oxo Conditions

A. OLEFINS

As stated in Section II above, the hydrogenation of a simple C_n olefin is thermodynamically favored [110] over hydroformylation. However, at 100°–150°C in the presence of dicobalt octacarbonyl catalyst, hydroformylation of olefins is the major reaction. The nature of the olefin, however, plays an important role in determining the course of the reaction. The presence of a tertiary carbon atom in the olefin promotes hydrogenation [124] of the double bond at the expense of hydroformylation. Hydroformylation of internally branched olefins [16] takes place to the least extent because of steric effects. In these cases little or no carbon monoxide insertion takes place and the hydrogenation of the double bond is the only reaction possible. Thus propylene gives a 75% yield of alcohol at 200°C in the presence of a mixture of two parts of hydrogen to one of carbon monoxide, at 300 atm pressure. Isobutylene under the same conditions yields only 35% alcohol and 53% saturated hydrocarbon. The course of oxo reaction and hydrogenation in the case of isobutylene is illustrated by reactions (105a) and (105b).

$$
\begin{array}{c}
\mathrm{H_3C} \\
\diagdown \\
\mathrm{H_3C}\diagup \mathrm{C{=}CH_2}
\end{array}
\xrightarrow{\mathrm{HCo(CO)_4}}
\begin{cases}
\dfrac{(\mathrm{H_3C})_2\mathrm{CHCH_2COCo(CO)_3}}{}\xrightarrow{\ \mathrm{H_2}\ }(\mathrm{H_3C})_2\mathrm{CHCH_2CHO}+\mathrm{HCo(CO)_3}\quad(105a)\\[2ex]
(\mathrm{H_3C})_2\mathrm{CHCH_2COCo(CO)_3}\xrightarrow{\ 2\,\mathrm{H_2}\ }(\mathrm{H_3C})_2\mathrm{CHCH_2CH_2OH}+\mathrm{HCo(CO)_3}\quad(105b)
\end{cases}
$$

Unsaturated ethers such as butyl vinyl ether, $C_4H_9OCH{=}CH_2$, and allyl ethyl ether, $C_2H_5OCH_2CH{=}CH_2$, when heated [133] with $HCo(CO)_4$ and carbon monoxide at a substrate-to-cobalt ratio of 5:1 gave both hydrogenation and hydroformylation products. The vinyl ether gave 10% reduction and 33% hydroformylation, whereas the allyl ether gave 29% reduction and 71% hydroformylation. The presence of an ether oxygen atom that can coordinate with cobalt may inhibit somewhat the insertion of carbon monoxide in these substances.

Olefins possessing double bonds conjugated with an unsaturated group, with another olefinic bond, or with a carbonyl group [133], are reduced preferentially to saturated aldehydes or ketones if the oxo reaction is carried out below 125°C. They are converted to saturated alcohols at 180°–200°C under a 150–200 atm pressure with a 2:1 molar ratio of hydrogen to carbon monoxide. Delocalization of the π electrons with subsequent reduction of double-bond character seems to be an important factor in the hydrogenation of α,β-unsaturated substances. This is exemplified by the difference in behavior of crotonic acid and crotonaldehyde; the latter compound, with 2.4 kcal/mole higher resonance energy than that of the former, is largely hydrogenated. Furan [106] yields both hydrogenation and hydroformylation products, whereas thiophene and phenanthrene with resonance energies greater than that of furan are exclusively hydrogenated. Polynuclear hydrocarbons such as anthracene, naphthacene, perylene, and fluoranthrene are reduced to the corresponding hydrogenated products [74]. Either one or two double bonds become hydrogenated in these cases.

Goetz and Orchin [133] have proposed reaction sequence (106) for the reduction of α,β-unsaturated aldehydes and ketones. Formation of a π

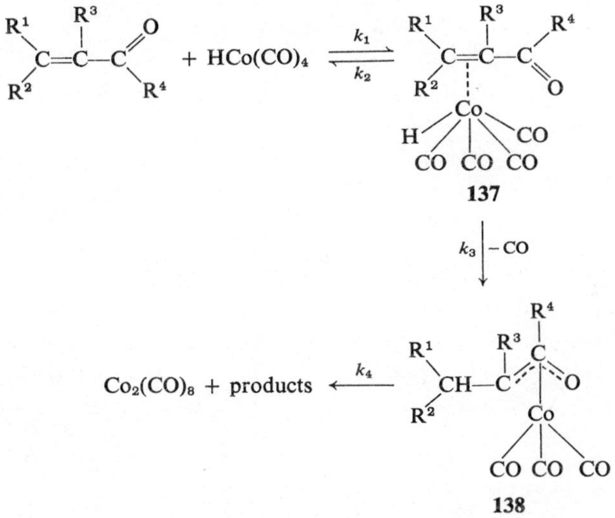

$$(106)$$

complex, **137**, is envisaged in the first step of the reaction. This complex then rearranges with the elimination of carbon monoxide and addition of hydrogen to form the π-oxapropenyl complex **138**, which is considered to be similar in structure to the well-known π-allyl complexes of cobalt. The rate of formation of complex **138** should decrease with increasing excess of carbon monoxide, and this has been verified experimentally [133]. The ketone 4-methyl-3-penten-2-one, when heated with a 5:1 molar ratio of $HCo(CO)_4$ under argon, underwent hydrogenation with a rate constant of 0.12 liter/mole minute whereas under carbon monoxide the rate was 0.032 liter/mole minute. Methyl substituents on the oxapropenyl complex should stabilize the complex and should give a higher rate of hydrogenation. This idea is supported by a higher yield of the saturated ketone when R^4 is CH_3 compared to the aldehyde obtained when R^4 is H. There is no direct proof for the formation of the π-oxapropenyl complex in this reaction, although indirect evidence is provided by the observations described above.

Heck [93] has reported that reactions of alkyl- and acylcobalt tetracarbonyl with α,β-unsaturated aldehydes and ketones give rise to the 1-acyloxy-π-

allylcobalt tricarbonyl complex **139** rather than 1-acylmethyl-π-oxapropenyl-cobalt tricarbonyl, **140**. This conclusion is based on infrared spectra of 1-acyloxy-π-allylcobalt tricarbonyls and their nonreactivity with bases.

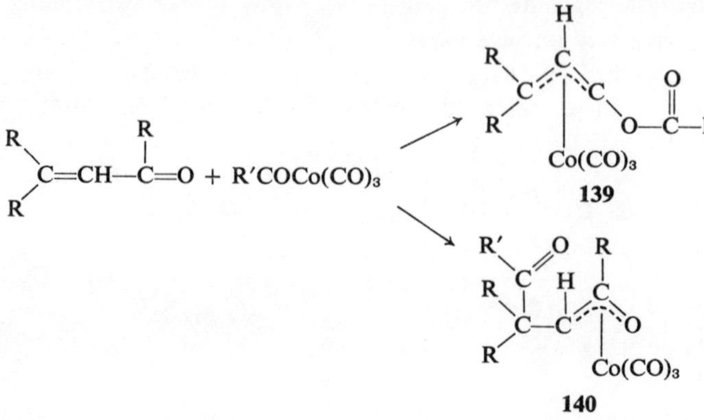

$$
\begin{array}{c}
R \\
\diagdown \\
C{=}CH{-}C{=}O + R'COCo(CO)_3 \\
\diagup \quad | \\
R \qquad R
\end{array}
\left\langle
\begin{array}{l}
\nearrow \quad \textbf{139} \\
\\
\searrow \quad \textbf{140}
\end{array}
\right.
\tag{107}
$$

Heck [93] has accordingly proposed that the reduction of α,β-unsaturated aldehydes and ketones in the presence of $HCo(CO)_4$ as catalyst proceeds by 1-hydroxy-π-allylcobalt tricarbonyl **142** rather than the π-oxapropenylcobalt tricarbonyl as suggested by Goetz and Orchin [133]. Heck's suggestion may be represented by the mechanism shown in (108). The π complex **141** probably

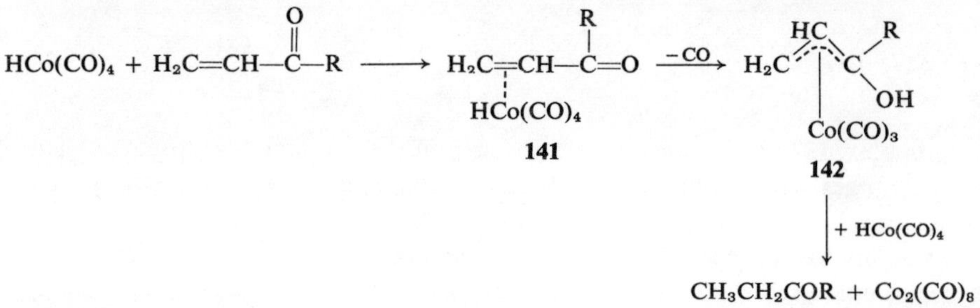

$$
HCo(CO)_4 + H_2C{=}CH{-}\overset{O}{\overset{\|}{C}}{-}R \longrightarrow
\underset{HCo(CO)_4}{\overset{\overset{R}{|}}{H_2C{=}CH{-}C{=}O}}
\xrightarrow{-CO}
\underset{Co(CO)_3}{\overset{HC\diagup^R\!\diagdown C}{H_2C\quad OH}}
\tag{108}
$$

$$
\textbf{141} \qquad\qquad\qquad \textbf{142}
$$

$$
\downarrow + HCo(CO)_4
$$

$$
CH_3CH_2COR + Co_2(CO)_8
$$

forms in the first step, as in Goetz and Orchin's [133] mechanism. Transfer of hydrogen to the carbonyl group of **141** then takes place, and with the loss of 1 mole of carbon monoxide, 1-hydroxy-π-allylcobalt tricarbonyl, **142**, is obtained. Reduction of **142** with $HCo(CO)_4$ and subsequent tautomerization yields the saturated aldehyde or ketone and $Co_2(CO)_8$. The catalyst $HCo(CO)_4$ may then be regenerated by the reaction of hydrogen with $Co_2(CO)_8$.

B. ALDEHYDES AND KETONES

One-step hydrogenation of aldehydes or ketones takes place when the oxo reaction is conducted at 180°–200°C and at 120–200 atm of a gaseous mixture of hydrogen and carbon monoxide. Butanal is reduced to 1-butanol, and hexanal is reduced to 1-hexanol under these conditions [106]. Iron penta-carbonyl, $Fe(CO)_5$, is more stable than $Co_2(CO)_8$ at higher temperatures and thus is an ideal catalyst [99] for hydrogenation at about 200°C. Hydrogena-tion of the substrate can be effected under much milder conditions by the use of tris(triphenylphosphine)–Rh(I) as the catalyst [77] with a 2:1 molar ratio of hydrogen to carbon monoxide at 50 atm pressure.

Wender et al. [134] provided evidence for the homogeneous nature of hydrogenation reactions catalyzed by $Co_2(CO)_8$ under oxo conditions. The basis of his work was the fact that a solution of $Co_2(CO)_8$ or $HCo(CO)_4$ in the alkene (e.g., propylene) and not the metallic phase, is the true catalyst for hydrogenation. Starting with metallic cobalt, at a partial pressure of hydrogen of 200 lb/inch², heterogeneous hydrogenation proceeded smoothly to give propane in 77% yield. When a small quantity of carbon monoxide at a partial pressure of 300 atm was introduced into the reaction mixture, the solid cobalt catalyst was seriously poisoned and no hydrogenation could be achieved. However, at a partial pressure of carbon monoxide of about 1000 lb/inch², approximately 1 mole of carbon monoxide per mole of alkene reacted and a 66% yield of 1-butanol was obtained. Evidently the solid metallic cobalt catalyst was converted into the carbonyl, which then func-tioned as the catalyst for hydrogenation. Thus under oxo conditions, at 300 lb/inch² of a 2:1 mixture of hydrogen and carbon monoxide, the reaction proceeds mainly by the homogeneous route with $Co_2(CO)_8$ or $HCo(CO)_4$ functioning as the catalyst. This interpretation of the change in reaction mechanism at high carbon monoxide pressure is also supported by the fact that sulfur compounds had no effect on the rate of hydrogenation, and thiophene carboxaldehyde or thiophene did not poison the catalyst. These compounds are notoriously effective in the poisoning of heterogeneous catalytic metal surfaces.

Support for the homogeneous hydrogenation mechanism has also been provided by Goetz and Orchin [135] and by Marko [96]. Aldridge et al. [136,137], however, provided evidence showing that part of the hydrogenation of aldehydes proceeds by a heterogeneous path involving metallic cobalt which is continuously consumed and regenerated during the reaction. This conclusion concerning the partial heterogeneous character of the hydrogena-tion reaction was also supported by the fact that the catalyst was poisoned by heavy metal ions of lead(II), mercury(II), and bismuth(II). These ions are readily reduced by the metallic cobalt to the corresponding metal. The thallium(III) ion, which cannot be reduced to metallic thallium by cobalt, has no adverse effect on the rate [137]. Thiophene retards the reaction by 25% at 160°C under 950 lb/inch² of carbon monoxide and 750 lb/inch² of hydrogen pressure. The observed variation of the retarding effect of thiophene with variation in carbon monoxide pressure rules out any competition of thiophene with aldehyde for the cobalt carbonyl species and is consistent with the poisoning of metallic cobalt, which is in dynamic equilibrium with $Co_2(CO)_8$. The rate of the hydrogenation is first order with respect to the aldehyde concentration in the range 239–953 $lb/inch^2$ of carbon monoxide at 160°C. The apparent rate constant of the hydrogenation reaction was found to be inversely proportional to the first power of carbon monoxide pressure in a manner similar to that reported [113] for the oxo reaction. The mechanism given in Eqs. (76) and (109)–(112) was proposed for hydrogenation of aldehydes. The first reaction is heterogeneous and results in the formation of

$$Co_2(CO)_8 + H_2 \underset{k_{-1}}{\overset{k_1}{\rightleftarrows}} 2\,HCo(CO)_4 \tag{76}$$

$$RCHO + HCo(CO)_4 \underset{k_4}{\overset{k_3}{\rightleftarrows}} \underset{HCo(CO)_3}{\overset{\overset{\displaystyle H}{|}}{RC}}{=}O + CO \tag{109}$$

$$\underset{HCo(CO)_3}{\overset{\overset{\displaystyle H}{|}}{RC}}{=}O \xrightarrow{k_5} \underset{H}{\overset{\overset{\displaystyle OH}{|}}{R{-}C{-}Co(CO)_3}} \tag{110}$$

$$\underset{H}{\overset{\overset{\displaystyle OH}{|}}{R{-}C{-}Co(CO)_3}} + HCo(CO)_4 \xrightarrow{k_6} RCH_2OH + Co_2(CO)_7 \tag{111}$$

$$Co_2(CO)_7 + CO \underset{k_8}{\overset{k_7}{\rightleftarrows}} Co_2(CO)_8 \tag{112}$$

$HCo(CO)_4$ from $Co_2(CO)_8$ with metallic cobalt as a catalyst. The hydrocarbonyl formed then acts as the homogeneous catalyst in the reduction of the aldehyde. With the third reaction (k_5) as the rate-determining step, the rate equation is given in (113).

$$\frac{d[RCH_2OH]}{dt} = \frac{k_3k_5[RCHO][HCo(CO)_4]}{k_4[CO]} \tag{113}$$

Formation of an aldehyde complex in reaction (109) and the subsequent addition of hydrogen to the carbonyl oxygen in the rate-determining step, is in accord with the scheme suggested by Heck [93] for the reduction of α,β-unsaturated aldehydes and ketones with $HCo(CO)_4$. Reactions (110) and (111) are similar to the mechanism of Goetz and Orchin [133] for the reaction of a π-oxapropenyl complex with $HCo(CO)_4$. Except for the first step involving the formation of cobalt hydrocarbonyl with metallic cobalt as a catalyst, the reactions are homogeneous.

Marko [96] reported that the rate of hydrogenation of propionaldehyde in the range 32–210 atm of carbon monoxide is inversely proportional to the square of the carbon monoxide pressure. At pressures lower than 32 atm of carbon monoxide, the reaction is heterogeneous and metallic cobalt is precipitated. To account for the inverse dependence of the reaction rate on carbon monoxide pressure, reaction sequence (114) was proposed.

$$\text{HCo(CO)}_3 \underset{-CO}{\overset{+CO}{\rightleftharpoons}} \text{HCo(CO)}_4 \underset{-CO}{\overset{+CO}{\rightleftharpoons}} \underset{\textbf{143}}{\text{HCOCo(CO)}_4} \tag{114}$$

Complex **143** is similar to the acylcobalt tetracarbonyl of Breslow and Heck [122] with a hydrogen atom in place of an alkyl group. Its formation, however, is not substantiated by evidence other than the observed reaction kinetics [96].

C. REDUCTION AND HOMOLOGIZATION OF ALCOHOLS

Wender *et al.* [134,138] have reported that aliphatic primary, secondary, and tertiary alcohols, when treated with cobalt octacarbonyl catalyst and oxo synthesis gas at 185°–190°C, are converted to homologous alcohols having one additional carbon atom. *t*-Butyl alcohol reacts at 150°–180°C with synthesis gas under pressure, with $Co_2(CO)_8$ as catalyst, to give isoamyl alcohol and a small quantity of neopentyl alcohol, as shown in Eq. (115).

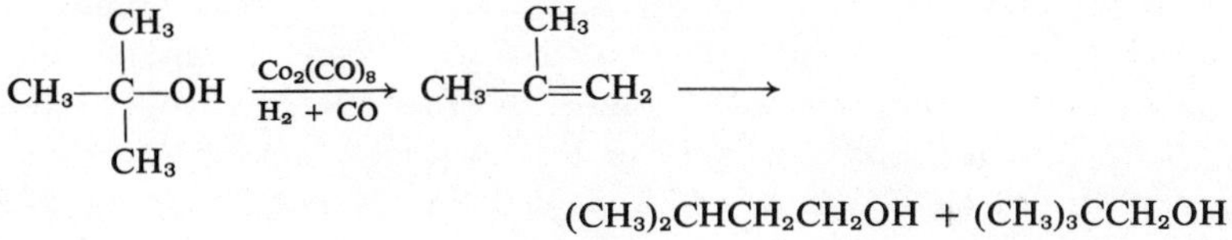

$$\underset{\substack{|\\ CH_3}}{\overset{\substack{CH_3\\ |}}{CH_3-C-OH}} \xrightarrow[\text{H}_2 + CO]{Co_2(CO)_8} \underset{}{\overset{\substack{CH_3\\ |}}{CH_3-C=CH_2}} \longrightarrow$$

$$(CH_3)_2CHCH_2CH_2OH + (CH_3)_3CCH_2OH \tag{115}$$

The reaction is visualized as proceeding through dehydration of *t*-butyl alcohol to isobutylene [90] (identified in the exhaust gas), which undergoes hydroformylation and reduction to the saturated alcohols. If the reaction is conducted below 150°C only isovaleraldehyde is obtained. Isopropyl alcohol reacts slowly to give an 11% yield of a mixture of isobutyl alcohol and *n*-butyl alcohols. Methyl alcohol gave a 39% yield of ethyl alcohol under the same conditions [74].

The mechanism of homologization in the case of methyl alcohol should be different from reaction sequence (115) since methyl alcohol cannot give an olefin. In this case methylcobalt tetracarbonyl may be formed. This substance may then rearrange to acetylcobalt tetracarbonyl and is finally converted to ethyl alcohol by reduction with hydrogen, in accordance with the reactions (116)–(119). Cobalt hydrocarbonyl is a strong acid and as such can readily form methylcobalt tetracarbonyl from methyl alcohol.

$$CH_3OH + HCo(CO)_4 \longrightarrow CH_3Co(CO)_4 + H_2O \qquad (116)$$

$$CH_3Co(CO)_4 \rightleftharpoons CH_3COCo(CO)_3 \overset{CO}{\rightleftharpoons} CH_3COCo(CO)_4 \qquad (117)$$

$$CH_3COCo(CO)_4 + H_2 \longrightarrow CH_3CHO + HCo(CO)_4 \qquad (118)$$

$$CH_3CHO + H_2 \longrightarrow CH_3CH_2OH \qquad (119)$$

A reaction sequence similar to the above is the acid-catalyzed rearrangement [139] of pinacol with $HCo(CO)_4$ to give an acylcobalt intermediate, which undergoes further reduction to yield pinacolyl alcohol.

Aromatic alcohols undergo homologization and reduction when hydrogenated under oxo conditions, as indicated by reaction (120). Homologization

$$C_6H_5CH_2OH \nearrow^{\displaystyle C_6H_5CH_2CHO \longrightarrow C_6H_5CH_2CH_2OH}_{\displaystyle \searrow C_6H_5CH_3 + H_2O} \qquad (120)$$

and reduction reactions take place by the direct interaction of the alcohol. Carbon atoms carrying more than one phenyl substituent give reduced products (hydrocarbons) exclusively and homologous alcohols are not obtained. Thus from benzhydrol and triphenylcarbinol [108] more than 95% yields of diphenylmethane and triphenylmethane, respectively, are formed.

The effect of adding pyridine on the rate of reduction of aromatic alcohols is the same as that observed in the oxo reaction [16]. Small amounts of pyridine in benzene (3% by volume) increase the rate, whereas larger amounts, 20–50%, decrease the rate. The activating effect of substituents [140] on the rate of reduction of benzyl alcohols to the corresponding hydrocarbons decreases in the order p-$CH_3O \gg p$-$CH_3 > H > p$-$Cl > m$-$OCH_3 \gg m$-$CF_3 \approx$ 2,4-diCl. This effect of substituents on the benzene ring in influencing the rate of reduction is not clearly understood. The effect of solvents [74] on the specific reaction rate constants for hydrogenation is in the order: acetone > methanol > ethanol > chlorobenzene > ethyl ether > butyl ether > benzene. It was postulated that homologization and reduction may proceed via a common intermediate aralkyl (benzyl) cobalt tetracarbonyl, formula 144, Eq. (121), which is converted to final products at different rates. Homologization may proceed through a reaction sequence similar to that illustrated above for methanol, reactions (117)–(119). Hydrogenation may occur by the interaction of an aralkylcobalt tricarbonyl with molecular hydrogen as proposed

$$C_6H_5CH_2OH + HCo(CO)_4 \longrightarrow C_6H_5CH_2Co(CO)_4 + H_2O \qquad (121)$$
$$\textbf{144}$$
$$C_6H_5CH_2Co(CO)_4 \longrightarrow C_6H_5CH_2Co(CO)_3 + CO \qquad (122)$$

$$C_6H_5CH_2Co(CO)_3 + H_2 \longrightarrow C_6H_5CH_3 + HCo(CO)_3 \qquad (123)$$

by Marko [96] [Eq. (123)]. Unlike the reactions with methanol described above, carbon monoxide insertion does not take place in the formation of the aralkylcobalt carbonyl and the only reaction possible for this compound is reduction by hydrogen to form the corresponding hydrocarbon, regenerating the catalyst.

A hydrogenation reaction that merits special consideration is the addition of hydrogen and hydroxymethyl groups across the double bond of an olefin, referred to as hydroxymethylation [reaction (124)]. This reaction proceeds [90] with iron pentacarbonyl as the source of the catalyst in the presence of an aqueous base rather than in an organic solvent. Water is the source of hydrogen and the products of reaction are straight-chain alcohols. Under the

$$H_2Fe(CO)_4 + 2\ CH_2{=}CH_2 + 4H_2O \longrightarrow 2\ CH_3CH_2CH_2OH + Fe(HCO_3)_2 \quad (124)$$

reaction conditions employed the ferrous bicarbonate is reduced with regeneration of the catalyst. The reaction can be conducted with carbon monoxide at elevated pressures and catalytic amounts of $Fe(CO)_5$ in alkaline solution. Under these conditions the overall reaction may be represented by Eq. (125).

$$CH_2{=}CH_2 + 3\ CO + 2\ H_2O \longrightarrow CH_3CH_2CH_2OH + 2\ CO_2 \quad (125)$$

V. Applications of the Oxo Reaction

The commercial potential of the oxo reaction may be estimated from the fact that oxo production capacity had grown [140a] to 5.7 billion pounds in 1969 from a small beginning of 5 million pounds in 1948. About 70% of the oxo production capacity is utilized for the synthesis of C_8, C_{10}, and C_{13} alcohols, and most of the remainder is involved in C_4 aldehyde production [141]. The C_8, C_{10}, and C_{13} alcohols are generally used as plasticizers and for the production of detergents and lubricants. The economic production of olefins by a recently developed cyclic cracking process [142] and by Zeigler catalysis [143] provides inexpensive raw materials for the one-step oxo synthesis of alcohols. In addition to alcohols, many secondary products, as indicated in Table II, may be obtained by modification of the primary oxo reaction. A reaction of great potential is the formation of dialdehydes and dialcohols from diolefins by the utilization of rhodium catalysts. These reactions have the potential of developing extensive commercial applications, and large markets for some of the products [16] obtained in this manner are already established. Conversion of olefins to long-chain monoaldehydes is another possible avenue for the application of "oxo chemistry" in the development of new commercial chemicals [144].

4

Hydrosilation of Alkenes and Alkynes

I. Catalysis by Platinum(II) and Rhodium(II)

"Hydrosilation" is the term usually employed for the addition of silicon hydrides to olefins in the presence of complexes of platinum, rhodium, and other metals as catalysts. This reaction takes place through an insertion mechanism in the coordination sphere of the metal ion, and is similar to insertion reactions such as catalytic hydrogenation and the oxo reaction. Addition of a number of silicon hydrides such as Cl_3SiH, CH_3Cl_2SiH, $(CH_3)_2ClSiH$, $C_6H_5Cl_2SiH$, and $(C_2H_5O)(CH_3)_2SiH$ to 1-pentene and 2-pentene to form n-alkylsilyl derivatives as indicated in reaction (126) was reported by Speier et al. [145]. The reaction takes place in the temperature

$$R^1R^2R^3SiH + RCH{=}CHR' \xrightarrow{\text{catalyst}} \begin{matrix} R^1R^2R^3Si & H \\ \diagdown & \diagup \\ R{-}C{-}C{-}R' \\ \diagup & \diagdown \\ H & H \end{matrix} \qquad (126)$$

range of 30°–100°C and is catalyzed by a variety of Group VIII metal ions, including those of platinum, rhodium, and iridium.

A. Isomerization of Alkenes

Hydrosilation of alkenes in most cases is accompanied by migration of the double bond. In the presence of chloroplatinic acid as catalyst, the hydrides Cl_3SiH, CH_3Cl_2SiH, or $(CH_3)_2ClSiH$ always form n-pentylchlorosilanes from 2-pentene, and n-heptylchlorosilanes from 2-heptene [146]. Migration of double bonds in the addition of silanes to olefins was reported by Saam and

Speier [147]. In the case of the α-olefins, 2-methyl-1-butene and 3-methyl-1-butene, the addition of the silicon moiety to the double bond takes place exclusively at the terminal carbon on the alkyl chain. The internal olefin,

$$
CH_3-CH_2-\overset{\overset{\displaystyle CH_3}{|}}{C}=CH_2 \xrightarrow{CH_3Cl_2SiH} CH_3Cl_2Si-CH_2-\overset{\overset{\displaystyle CH_3}{|}}{CH}-CH_2-CH_3 \quad (127)
$$
$$
\textbf{145}
$$

$$
CH_2=CH-\overset{\overset{\displaystyle CH_3}{|}}{HC}-CH_3 \xrightarrow{CH_3Cl_2SiH} CH_3Cl_2Si-CH_2-CH_2-\overset{\overset{\displaystyle CH_3}{|}}{HC}-CH_3 \quad (128)
$$
$$
\textbf{146}
$$

2-ethyl-2-butene, when heated with H_2PtCl_6 and CH_3Cl_2SiH gives 30% **145** and 70% **146**.

$$
CH_3-CH=\overset{\overset{\displaystyle CH_3}{|}}{C}-CH_3 \xrightarrow[Pt(II)]{CH_3Cl_2SiH}
\begin{cases}
CH_3Cl_2Si-CH_2-CH(CH_3)-CH_2-CH_3 & \textbf{145} \\
CH_3Cl_2Si-CH_2-CH_2-CH(CH_3)_2 & \textbf{146}
\end{cases} \quad (129)
$$

Another example of double-bond migration is the conversion of 1-methylcyclohexene to (cyclohexylmethyl)methyldichlorosilane, **147**, and (3-methylcyclohexyl)methyldichlorosilane, **148**. The hydrosilation reactions (129) and

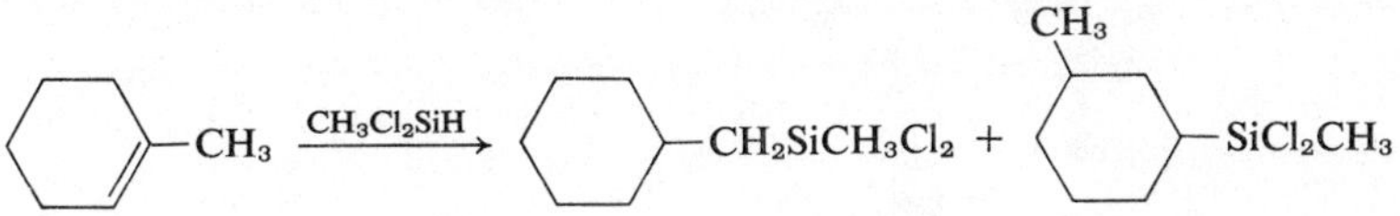

$$
\quad (130)
$$

147   **148**

(130) are very rapid even in the presence of such low catalyst concentrations as 10^{-8} mole of catalyst per mole of reactant [146]. The nature of the intermediates in these reactions has not been elucidated.

The addition of trimethylsilane at 45°C to 1,5- and 1,3-cyclooctadiene takes place in the presence of any platinum catalyst; for example, H_2PtCl_6, *trans*-dichloro(ethylene)(pyridine)platinum(II), 5% Pt metal on charcoal, etc., gave exclusively 3-trimethylsilylcyclooctene, **149**, as indicated in Eq. (131). A mixture of 1,5-cyclooctadiene and trimethylsilane when heated at 45°C for 36 hours in the presence of H_2PtCl_6 gave 3-trimethylsilylcyclooctene in 35%

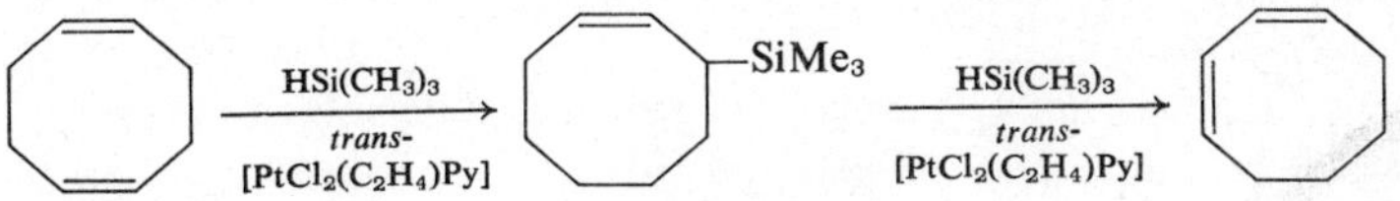

$$
\quad (131)
$$

149

yield [148]. The use of *trans*-dichloro(ethylene)(pyridine)platinum(II) in benzene as catalyst resulted in a 55% yield of 3-trimethylsilylcyclooctene from 1,5-cyclooctadiene. It has been proposed [148] that rapid isomerization of 1,5-cyclooctadiene to the 1,3-isomer takes place through the influence of the platinum(II) followed by 1,4-addition of the silane to the conjugated diene.

Chalk and Harrod [149] have studied hydrosilation and isomerization of olefins with platinum(II), rhodium(I), and iridium(I) catalysts. In order to study the nature of the intermediate species involved, reactions of tertiary phosphine complexes of platinum(II) and iridium(I) with silanes were investigated [149]. In the case of 1-hexene, the extent of isomerization that takes place during the hydrosilation reaction depends on the nature of the silane used. Trimethoxy- and triethoxysilanes caused no isomerization of 1-hexene, whereas silanes with electronegative substituents such as trichloroethyl, dichloro, and phenyldichloro caused extensive isomerization of the excess olefin present in the reaction mixture. An exceptional feature of the isomerization reaction is the constant ratio of 0.4:1 of the *cis*-2- to *trans*-2-isomer throughout the course of the reaction when phenyldichlorosilane was used as the substrate. The amount of *cis*-2-isomer formed does not approach its equilibrium value during the course of hydrosilation; also the reaction does not show an initial preference for the formation of the *cis*-2-isomer. The ratio of *cis*-2- to *trans*-2-isomers, however, does vary somewhat with the nature of the silane. For ethyldichloro- and trichlorosilane, the ratios were found to be 0.53:1 and 0.31:1, respectively.

B. THE HYDROSILATION REACTION

The reaction mixture of bis(tributylphosphine)dichloroplatinum(II) and triethylsilane when refluxed under nitrogen for 1 hour results in the rapid exchange of hydride and chloride ions to give a chlorohydridoplatinum(II), (Pt—H stretching frequency, $2150 \pm 10 \, \text{cm}^{-1}$) and the corresponding triethylchlorosilane [149], as indicated in reaction (132). Bis(triphenylphosphine)-

$$[Bu_3P]_2Cl_2Pt + (C_2H_5)_3SiH \longrightarrow [Bu_3P]_2HClPt + (C_2H_5)_3SiCl \qquad (132)$$

carbonylchloroiridium(I), **150**, reacts with trialkyl- and trialkoxysilanes in a different manner. Triethoxysilane reacts with [PPh$_3$]$_2$IrCOCl complex, **150**, at room temperature to form an adduct of iridium(III), **151**, that retains both the silicon and the hydride fragments. The adduct **151** was characterized by strong infrared bands at 1970–2040 cm^{-1}, assigned to the stretching frequency of the iridium(III)–carbonyl group. Bands in the 2080–2180 cm^{-1} region were assigned to the Ir—H stretching frequency of iridium(III) hydride. The reaction thus involves homolytic splitting of the silicon–

hydrogen bond. A similar reaction of complex **150** with molecular hydrogen results in the formation of the dihydride **152**.

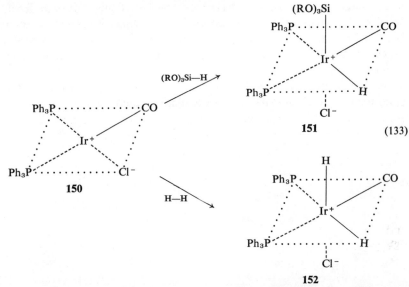

(133)

The reaction of bis(triphenylphosphine)dichloroplatinum(II), **153**, with trialkylsilanes is illustrated in reaction (134) as taking place in a manner analogous to reaction (133). The platinum(IV) complex **154** is unstable and readily reverts to the planar platinum complex **155**, with elimination of chlorosilane [Eq. (134)].

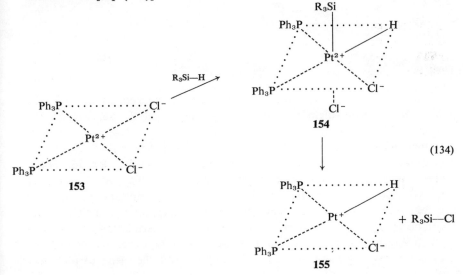

(134)

Catalysis of hydrosilation of 1-hexene was studied with tetrakis(ethylene)-μ'-dichlorodirhodium(I) and dichlorobis(ethylene)-μ,μ-dichlorodiplatinum(II) Both platinum(II) and rhodium(I) π complexes of this type were found to be very effective as catalysts for the addition of olefins to silanes. An interesting fact indicating similar reaction mechanisms is that rates, yields, and product distributions were identical with both platinum(II) and rhodium(I)–olefin complexes.

The mechanism given in Eq. (135) was suggested by Chalk and Harrod [149] for the metal ion-catalyzed hydrosilation reaction.

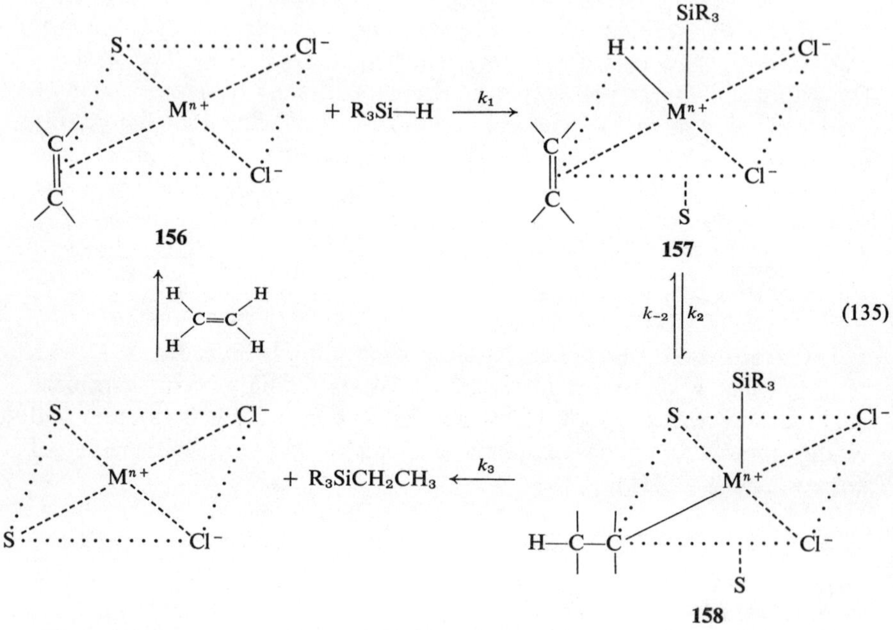

(135)

(S = solvent)

The first step in the mechanism involves the homogeneous splitting of the silane by the metal–olefin complex to form an octahedral oxidative addition product, **157**, containing both silyl and hydrido ligands. Metal ions with a d^8 configuration [e.g., platinum(II), rhodium(I), and iridium(I)] are capable of this type of reaction. In cases where the olefin–hydrido complex, **157**, is not very stable, two-electron reduction of the metal takes place with the release of hydrogen ions. In the case of palladium(II)–olefin complexes, the metal ion is reduced to the free metal. Since platinum(II) salts and platinum(II)–olefin complexes are equally effective as catalysts in hydrosilation reactions, a kinetically equivalent mechanism would consist of homogeneous splitting of the silane by platinum(II) followed by addition of the olefin to form a π

complex [Eqs. (136) and (137)]. Since the Si—H bond is a relatively weak homoplanar bond the homogeneous splitting of the Si—H bond in the first step of hydrosilation seems entirely reasonable. This concept is further supported by the isolation by Chalk and Harrod [149] of an octahedral iridium complex containing silyl and hydrogen ligands.

$$PtS_2X_2 + R_3SiH \longrightarrow PtHS_2X_2(SiR_3) \tag{136}$$

$$PtHS_2X_2(SiR_3) + olefin \longrightarrow (olefin)PtHSX_2(SiR_3) \tag{137}$$

The reversible equilibrium illustrated in Eq. (135) between silyl metal–olefin and silyl metal–alkyl complexes leads to isomerization of the olefin [149]. In cases where k_{-2} becomes significant relative to k_3, isomerization of the olefin takes place. The isomerization of the olefin in hydrosilation thus depends on the competition between the reversible metal–olefin, metal alkyl equilibrium with other possible reactions of the complex intermediates. Complex 158 may eliminate the alkylsilane moiety and reform to metal–olefin complex 156 on reaction with an additional olefin molecule. In passing from complex 158 to complex 156 a change in bonding and configuration of the metal complex from octahedral to a planar tetracoordinate complex takes

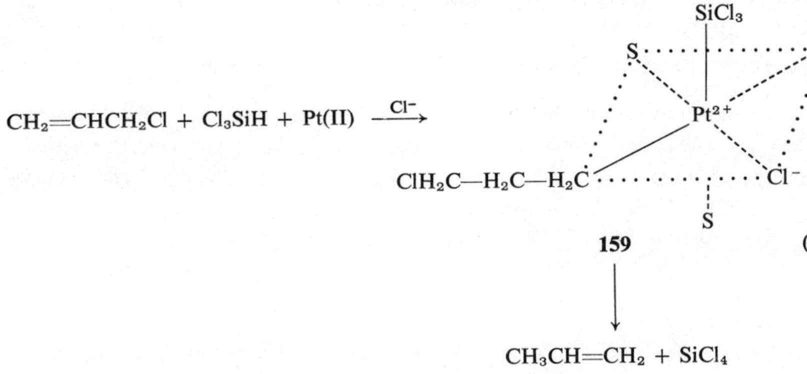

$$CH_2{=}CHCH_2Cl + Cl_3SiH + Pt(II) \xrightarrow{Cl^-}$$

159 (138)

$$CH_3CH{=}CH_2 + SiCl_4$$

place. Chelate π complexes of platinum(II) and rhodium(I) with diolefins show considerably less reactivity than do the complexes of monoolefins as hydrosilation catalysts, presumably because of the greater stabilities of the π complexes. Tertiary phosphine complexes of platinum(II) that form very unstable metal–olefin complexes can take part only in ligand-exchange reactions with silanes and are inactive as hydrosilation catalysts. In the case of allyl chloride the alkyl metal complex, 159, eliminates silicon halide, $SiCl_4$, with the formation of propylene, as indicated in reaction (138).

Silylplatinum hydride complex intermediates have been proposed by Bank *et al.* [150] and by Ryan and Vetter [151] to explain deuterium–hydrogen exchange observed in the hydrosilation of isobutylene by deuterotrichlorosilane in the presence of H_2PtCl_6 as a catalyst. With excess deuterotrichlorosilane the reaction product, isobutyltrichlorosilane, was found to have an overall degree of deuteration of 70%. The unreacted silane isolated from the reaction mixture had absorption maxima corresponding to both Si—H and Si—D bonds, with absorptions at 2350 cm^{-1} and 1645 cm^{-1}, respectively, and with 40% of the material in the deuterated Cl_3SiD form. The mechanism illustrated by Eqs. (139)–(141) was proposed to explain hydrogen–deuterium exchange in deuterotrichlorosilane and trichlorosilane and formation of deuterated isobutyltrichlorosilane. In practice, additional exchange resulting

$$Cl_3SiD + PtCl_6{}^{2-} \; \rightleftharpoons \; Cl_3SiPtDCl_4{}^{2-} \; \xrightleftharpoons{Cl_3SiD}$$
$$(Cl_3Si)_2PtD_2Cl_2{}^{2-} \quad (139)$$

$$(Cl_3Si)_2PtD_2Cl_2{}^{2-} + 2\;CH_2{=}C(CH_3)_2 \; \rightleftharpoons \; [Cl_3SiCH_2C(CH_3)_2]_2PtD_2Cl_2{}^{2-} \quad (140)$$

$$[Cl_3SiCH_2C(CH_3)_2]_2PtD_2Cl_2{}^{2-} + Cl_3SiD \; \rightleftharpoons$$
$$Cl_3SiC_4H_7D_2 + HSiCl_3 + Cl_3SiPtHDCl_3{}^{2-} \quad (141)$$

from continued equilibrium processes resulted in a silation product containing additional label approximating $Cl_3SiC_4H_{6.5}D_{2.5}$.

C. HYDROSILATION OF ALKYNES

The catalysis of homogeneous hydrosilation of alkynes by d^8 metal ions has not yet been reported. Benkeser and Hickner [152] have reported the heterogeneous platinum/charcoal-catalyzed hydrosilation of 3,3-dimethyl-1-butyne with trichlorosilane. The addition of trichlorosilane to the acetylene is *cis*, with the formation of *trans*-1-trichlorosilyl-3-methyl-1-butene **160**, as

$$\begin{array}{c} CH_3 \\ | \\ H_3C\text{—}C \\ H_3C{-}C{-}C{\equiv}C{-}H \xrightarrow[\text{Pt-C}]{Cl_3SiH} H_3C \quad \diagdown \qquad \diagup H \\ H_3C \qquad\qquad\qquad\qquad H \quad C{=}C \\ \qquad\qquad\qquad\qquad\qquad\qquad SiCl_3 \end{array} \quad (142)$$

160

illustrated in reaction (142). The hydrosilation reaction of this alkyne was also attempted with the use of chloroplatinic acid as a catalyst [153]. Chloroplatinic acid was partially reduced to metallic platinum during the reaction that resulted again in the formation of the *trans* product [153]. Further hydrosilation of the *trans* olefin **160** either by chloroplatinic acid or platinized charcoal was not successful. *cis*-1-Trichlorosilyl-3,3-dimethyl-1-butene, however, can be further hydrosilated to a saturated product [153].

II. Catalysis of Hydrosilation by Metal Carbonyls

Nesmayanov *et al.* [154] have found that silanes add to olefins in the presence of small quantities of iron pentacarbonyl at $100°-140°C$ to form both addition and substitution products, depending on the molar ratio of olefin to silane. For example, the reaction of triethylsilane with ethylene at a molar ratio of 3:1 forms the normal addition product, **161**, whereas at a ratio of 5:1 of silicon:olefin, **162** is obtained [reaction (143)]. The factors controlling the

$$(C_2H_5)_3SiH + CH_2{=}CH_2 \nearrow \begin{array}{c} (C_2H_5)_3SiCH_2CH_3 \\ \textbf{161} \end{array} \searrow \begin{array}{c} (C_2H_5)_3SiCH{=}CH_2 \\ \textbf{162} \end{array} \tag{143}$$

ratio of addition to substitution are as yet not clear and the mechanism of catalysis by iron pentacarbonyl has not been elucidated.

Chalk and Harrod [155] have reported the catalysis of hydrosilation by dicobalt octacarbonyl, $Co_2(CO)_8$, in the reaction of 1-octene with a variety of silanes, including $(CH_3O)_3SiH$, $(C_2H_5)_3SiH$, and $(C_6H_5)Cl_2SiH$ under milder conditions than those required for iron pentacarbonyl [154]. The reaction can be conducted in the range of $0°-60°C$ under ordinary pressure. In this case hydrosilation of olefins takes place with much more extensive isomerization than was observed for the platinum(II)- and rhodium(I)-catalyzed reactions [149]. The rate of isomerization is much faster than hydrosilation, and some of the isomeric olefins formed require a long time for completion of the hydrosilation reaction.

To explain the isomerization, conversion of the olefin to a σ-bonded alkyl group by hydride transfer from $HCo(CO)_4$ has been suggested [156], as indicated in Eq. (144). Reaction (144) is responsible for rapid isomerization of

$$HCo(CO)_4 + RCH{=}CH_2 \nearrow \begin{array}{c} RCH_2CH_2Co(CO)_4 \\ \end{array} \searrow \begin{array}{c} CH_3 \\ | \\ RCHCo(CO)_4 \end{array} \tag{144}$$

the olefins in these hydrosilation reactions. For 1-pentene, isomerization proceeds at a rate about twenty times that of the rate of hydrosilation [156]. The isomerization reaction is analogous to that proposed by Heck and Breslow [17] for the isomerization of olefins in the oxo reaction. Hydrosilation of the alkene is then effected by reaction of the alkylcobalt tetracarbonyl with the silane [156] [Eq. (145)]. Excess olefin favors hydrosilation since the above equilibrium between $HCo(CO)_4$ and the olefin illustrated in (144) is shifted toward the right at high olefin concentration. Further insight into the mechanism of metal carbonyl-catalyzed hydrosilation is provided by a study of the

$$RCH_2CH_2Co(CO)_4 + \;{\backslash}\!\!-\!\!Si\!-\!H \longrightarrow HCo(CO)_4 + RCH_2CH_2\!-\!Si\!\!-\!\!{\diagup} \quad (145)$$

interaction of cobalt octacarbonyl and silanes in a 1:1 ratio in the absence of the olefin [149,156]. The silicon–hydrogen bond is homolytically cleaved by $Co_2(CO)_8$ to form hydrocarbonyl and a silylcobalt derivative, as illustrated by Eq. (146).

$$Co_2(CO)_8 + \;{\backslash}\!\!-\!\!Si\!-\!H \longrightarrow \;{\backslash}\!\!-\!\!Si\!-\!Co(CO)_4 + HCo(CO)_4 \quad (146)$$

163

The reaction of dicobalt octacarbonyl with silanes is accompanied by the disappearance of IR absorption bands due to silicon–hydrogen stretch and the bands at 1860 cm^{-1} that are due to the bridging carbonyl groups. The silylcobalt complex **163** is relatively stable and reacts only slowly with air or moisture. Apparently the high stability is due in large measure to the formation of a bond with d–d_π character between the cobalt and silicon atoms. If a large excess of silane is added to dicobalt octacarbonyl, so that formation of the silylcobalt tetracarbonyl is complete, and the olefin then added to the reaction mixture, the hydrosilation reaction is completely inhibited. Also, at a 1:1 ratio of $Co_2(CO)_8$ to silane, catalytic activity is markedly diminished at temperatures above 60°C, because of the reaction of $HCo(CO)_4$ with the silane to form alkylcobalt tetracarbonyl and molecular hydrogen [156], as indicated in Eq. (147).

$$HCo(CO)_4 + R_3SiH \longrightarrow R_3SiCo(CO)_4 + H_2 \quad (147)$$

A. Deuterium-Exchange Studies

Sommer and Lyons [157] have investigated the catalysis of Si–H and Si–D exchange by $Co_2(CO)_8$ in homogeneous solution. The Si–H, Si–D exchange between α-naphthylmethylphenylsilane and phenylmethylethyldeuterosilane occurs with complete retention of configuration of the asymmetric silicon

$$\underset{\underset{CH_3}{|}}{\overset{\overset{C_6H_5}{|}}{\alpha\text{-}C_{10}H_7\!-\!Si\!-\!H}} + \underset{\underset{C_2H_5}{|}}{\overset{\overset{CH_3}{|}}{C_6H_5\!-\!Si\!-\!D}} \xrightarrow{Co_2(CO)_8} \underset{\underset{CH_3}{|}}{\overset{\overset{C_6H_5}{|}}{\alpha\text{-}C_{10}H_7\!-\!Si\!-\!D}} + \underset{\underset{C_2H_5}{|}}{\overset{\overset{CH_3}{|}}{C_6H_5\!-\!Si\!-\!H}} \quad (148)$$

centers. A silylcobalt tetracarbonyl formed according to reaction (149) has been proposed as an intermediate in the stereospecific Si–H and Si–D exchange: The α-naphthylphenylmethylsilylcobalt tetracarbonyl has been isolated as a white crystalline solid (m.p. 102°–104.5°C) and is itself a catalyst

$$2 \; \alpha\text{-}C_{10}H_7 \overset{\overset{\textstyle C_6H_5}{|}}{\underset{\underset{\textstyle CH_3}{|}}{Si}} H + Co_2(CO)_8 \longrightarrow 2 \; \alpha\text{-}C_{10}H_7 \overset{\overset{\textstyle C_6H_5}{|}}{\underset{\underset{\textstyle CH_3}{|}}{Si}} Co(CO)_4 + H_2 \quad (149)$$

for the exchange reaction. The overall reaction between α-naphthylphenyl-methylsilane and phenylmethylethyldeuterosilane thus seems to take place in two steps with complete retention of configuration at the asymmetric silicon centers.

$$\alpha\text{-}C_{10}H_7 \overset{\overset{\textstyle C_6H_5}{|}}{\underset{\underset{\textstyle CH_3}{|}}{Si}} Co(CO)_4 + C_6H_5 \overset{\overset{\textstyle CH_3}{|}}{\underset{\underset{\textstyle C_2H_5}{|}}{Si}} D \rightleftharpoons \alpha\text{-}C_{10}H_7 \overset{\overset{\textstyle C_6H_5}{|}}{\underset{\underset{\textstyle CH_3}{|}}{Si}} D + C_6H_5 \overset{\overset{\textstyle CH_3}{|}}{\underset{\underset{\textstyle C_2H_5}{|}}{Si}} Co(CO)_4$$

$$(150)$$

$$C_6H_5 \overset{\overset{\textstyle CH_3}{|}}{\underset{\underset{\textstyle C_2H_5}{|}}{Si}} Co(CO)_4 + \alpha\text{-}C_{10}H_7 \overset{\overset{\textstyle C_6H_5}{|}}{\underset{\underset{\textstyle CH_3}{|}}{Si}} H \rightleftharpoons C_6H_5 \overset{\overset{\textstyle CH_3}{|}}{\underset{\underset{\textstyle C_2H_5}{|}}{Si}} H + \alpha\text{-}C_{10}H_7 \overset{\overset{\textstyle C_6H_5}{|}}{\underset{\underset{\textstyle CH_3}{|}}{Si}} Co(CO)_4$$

$$(151)$$

Oxidation of Alkenes and Alkynes

The reactions discussed in this chapter represent two aspects of a general *cis*-ligand insertion reaction in which an alkene or alkyne is inserted between a metal and an —OH bond, followed by subsequent rearrangement. In the case of oxidation of alkenes, the metal ion catalyst participates in a two-electron oxidation–reduction cycle in the presence of suitable oxidizing agents. In the absence of a regenerating co-catalyst, however, the metal ion acts as a reagent and is reduced either to a lower valence state or to the free metal. Hydration of acetylene is not associated with the reduction of the metal ion, but is in effect the addition of hydrogen and hydroxyl groups to the triple bond, followed by rearrangement. The oxidation of olefins by a variety of metal ions [platinum(II), palladium(II), thallium(III), and lead(IV)] will be the subject matter of Section I. This will be followed by a consideration of the hydration of acetylenes in Section II.

I. Oxidation of Alkenes

Anderson [158] reported the hydration of the ethylene–platinum chloride complex $[C_2H_4PtCl_3]^-$ to yield acetaldehyde and metallic platinum. It was

$$C_2H_4PtCl_3^- + H_2O \longrightarrow CH_3CHO + 2\ HCl + Pt^0 + Cl^- \qquad (152)$$

proposed [159] that the oxidation of ethylene to acetaldehyde takes place through intermediate formation of ethyl alcohol, which is then oxidized to acetaldehyde by platinum(II). The concept of intermediate formation of ethyl alcohol in the oxidation of Zeise's salt was disputed by Joy and Orchin [160] on the basis of their studies on the hydration of this compound in the presence of various mineral acids. Direct formation of the aldehyde was proposed in accordance with the above equation but the kinetics of the reaction was not elucidated.

An analogous reaction, Eq. (153), is that of ethylene with rhodium trichloride to give acetaldehyde and a sparingly soluble rhodium(I) complex, $[(C_2H_4)_2RhCl]_2$ [160]. The oxidation of ethylene is thus accompanied by the reduction of rhodium(III) to rhodium(I), which forms a π complex with excess ethylene present in the reaction mixture.

$$6\ C_2H_4 + 2\ RhCl_3 + 2\ H_2O \longrightarrow$$

$$2\ CH_3CHO + (C_2H_4)_2Rh^+ \overset{Cl^-}{\underset{Cl^-}{\diamond}} {}^+Rh(C_2H_4)_2 + 4\ HCl \qquad (153)$$

Most of the interest in the oxidation of alkenes is centered on the use of a palladium(II) rather than a platinum(II) catalyst, since the π complexes of the former with alkenes are not as stable as those of the latter and are therefore more reactive. This difference may be the result of the greater extent of back π-bonding in platinum(II) as compared to the palladium(II) complexes. This makes the electron density around the olefinic double bond much less in palladium(II), thus facilitating [161] nucleophilic attack on the carbon–carbon double bond.

Another factor favoring catalytic activity of palladium complexes over that of platinum complexes is the more facile expansion of its coordination sphere to accommodate a fifth and sixth ligand [162]. This reduces the activation energy of nucleophilic addition since the nucleophile can be coordinated to the metal ion while attacking the carbon–carbon double bond.

Palladium(II) catalyzes the oxidation of ethylene to acetaldehyde [163–175], and substituted alkenes to aldehydes or ketones [103,169,176]. The presence of copper(II) as co-catalyst in the system oxidizes palladium(0) back to palladium(II) and keeps the process continuous. With copper(II) present, the oxidation of ethylene may be represented by the net reaction illustrated by Eq. (154). Reaction has been studied in detail kinetically [164,165] and the

$$C_2H_4 + \tfrac{1}{2}O_2 \xrightarrow[\text{CuCl}_2]{\text{PdCl}_2} C_2H_4O \qquad (154)$$

findings have been exploited industrially as Walker's process for the large-scale production of acetaldehyde from ethylene. The oxidation of unsaturated compounds with palladium(II) has grown in scope during the last few years and has recently been summarized by Stern [55].

The general reaction of an olefin with water and a salt of palladium(II) to form carbonyl compounds with the same number of carbon atoms as the original olefin may be represented by Eq. (155). The oxidation of an olefin by

$$RCH{=}CHR + H_2O + PdX_2 \longrightarrow RCH_2COR + Pd^0 + 2\,HX \qquad (155)$$

palladium(II) is restricted to those olefins having at least one hydrogen on each of the carbon atoms in the double bond. In the case of ethylene, acetaldehyde is the main product, as noted in Eq. (155). Propylene and higher α-olefins give a mixture of aldehydes and ketones. Internal olefins usually give ketones as the main product. Chain length, branching, and the nature of alkyl and aryl substituents on the olefin affect the ratio of ketones to aldehydes produced in a manner not yet completely elucidated [55]. The reaction takes place smoothly at room temperature and depends on the nature of the olefin used, the nature of the palladium(II) salt employed, the acidity of the medium, and the nature and concentration of the anions in solution.

The kinetics of the oxidation of ethylene by palladium(II) was reported by Smidt et al. [168,169,177], by Vargaftif et al. [171,172], and by Henry [163]. The oxidation of ethylene by palladium(II) is first order in ethylene and palladium(II) ion and is inhibited by chloride and hydrogen ions [172, 177, 178]. The oxidation of cyclohexene, however, was not inhibited by hydrogen ions [171]. There does not seem to be general agreement regarding the formation of an intermediate palladium(II)–olefin complex as the first preequilibrium step [103,167,168,163,178]. Such a complex may be a monomer or a dimer as reported originally by Kharasch et al. [180]. In cases where a mononuclear palladium(II)–ethylene complex is formed, the equilibria may be represented as Eqs. (156) and (157). Equation (156) is favored as the first

$$PdCl_4{}^{2-} + C_2H_4 \; \overset{K}{\rightleftharpoons} \; PdCl_3C_2H_4{}^- + Cl^- \qquad (156)$$

$$[PdCl_2 \cdot (OH)(H_2O)]^- + C_2H_4 \; \overset{K'}{\rightleftharpoons} \; [PdCl_2(OH)C_2H_4]^- + H_2O \qquad (157)$$

step in the reaction of ethylene with tetrachloropalladate(II) by Smidt et al. [177] and by Henry [175]. Henry [163] has reported the value of the equilibrium constant to be $K = 17.4 \pm 0.4$ at 25°C $[\mu = 2.0\ M\ (KClO_4)]$.

The complex formation constants K and K' have been found to decrease in the order ethylene > propylene > 1-butene > cis-2-butene > trans-2-butene [103,168,169], in good agreement with the relative values reported for complex formation constants of the same olefins with silver(I) [103]. The overall rate of hydration of the same olefin, however, is not influenced to any appreciable extent by the stability of the olefin complex, indicating that the small observed variation in reaction rate with the structure of the olefin is due to factors other than the influence of structure on the equilibrium formation constants such as K and K'.

Increase in temperature tends to reverse the equilibrium between ethylene and the palladium(II) species. At room temperature, high conversion to the ethylene–palladium(II) complex takes place, followed by slow hydrolysis, as indicated by Eq. (158). The rate constant of the overall reaction is the product

$$[PdCl_3C_2H_4]^- + H_2O \xrightarrow{k} C_2H_4O + Pd(0) + 3\,Cl^- + 2\,H^+ \qquad (158)$$

of the rate constant k and the equilibrium constant K for formation of the olefin complex (156). The temperature effect has been generally studied over a limited range of temperature and thus precludes any conclusions regarding relative sensitivity of the rate-determining and equilibrium steps to temperature changes.

The olefin hydration reaction is inhibited by strong acids [103, 163, 172, 177], the rate of the reaction having been found to be inversely proportional to the first power of the hydrogen ion concentration. At very low hydrogen ion concentration, however, the rate of the reaction decreases with decreasing hydrogen ion concentration and a plot of rate versus pH passes through a maximum [176, 178].

Halides decrease the rate of oxidation of olefins in the order chloride > bromide > iodide, which is the order in which they compete as ligands with the olefin in the coordination of platinum(II), thus shifting the equilibrium between $[PdCl_4]^{2-}$ and olefin away from complex formation. The rate of the reaction has been reported to be inverse second order with respect to chloride ion concentration, in the concentration range 0.1–0.3 M [103,163,167]. At very low chloride concentration, however, the rate of the reaction increases linearly with chloride ion concentration and passes through a maximum [176]. Anions such as sulfate, nitrate, perchlorate, and fluoride that form complexes with palladium(II) that are less stable than olefin complexes, have no retarding effect on the rate of reaction [179].

The effect of ionic strength on the rate of the reaction is difficult to interpret. For a sodium perchlorate medium, the rate increases with the ionic strength and reaches a maximum at $\mu = 0.4\,M$ and then decreases with increasing ionic strength [163]. The rate decreases above $\mu = 0.7\,M$ in $LiClO_4$ medium [172]. Since perchlorate ion does not complex palladium(II), the ionic medium seems to promote the dissociation of weak acids (as discussed below) and also influences to a certain extent the degree of aggregation of the reacting palladium(II) species in solution.

The rate law given in Eq. (159) is in accord with the first-order dependence of the reaction on olefin and palladium(II) concentrations and inverse first-order and second-order dependence on hydrogen and chloride ions, respectively, as reported by Henry [163].

$$-\frac{d[\text{olefin}]}{dt} = \frac{kK[PdCl_4{}^{2-}][C_2H_4]}{[Cl^-]^2[H^+]} \qquad (159)$$

where K is the equilibrium constant for the reaction (156) of C_2H_4 with $PdCl_4{}^{2-}$ and k is the rate constant of the hydrolysis of the olefin complex.

A general mechanism of the reaction consists of [103,178] (1) formation of olefin–Pd(II) π complexes in a series of preequilibrium steps; (2) rate-determining rearrangement of the π-olefin complex to a σ complex; (3) final breakdown of the σ complex to the products.

The preequilibrium steps in the case of ethylene may be represented as shown in Eqs. (160)–(162).

$$C_2H_4 + [PdCl_4]^{2-} \xrightleftharpoons{K_1} [C_2H_4PdCl_3]^- + Cl^- \qquad (160)$$
$$\mathbf{164}$$
$$[C_2H_4PdCl_3]^- + H_2O \xrightleftharpoons{K_2} [C_2H_4PdCl_2(H_2O)] + Cl^- \qquad (161)$$
$$\mathbf{165}$$
$$[C_2H_4PdCl_2(H_2O)] \xrightleftharpoons{K_3} [C_2H_4PdCl_2(OH)]^- + H_3O^+ \qquad (162)$$
$$\mathbf{166}$$

The formation of the π-C_2H_4–Pd(II) complexes **164** and **165** in steps (160) and (161) explain the inhibition by chloride. The aquo complex $[C_2H_4$-$PdCl_2(H_2O)]$, **165**, behaves as a weak acid and dissociates to give the hydroxo species, $[C_2H_4PdCl_2(OH)]^-$, **166**. The dissociation constant corresponding to equilibrium (162) has been estimated as 10^{-6} [163]. Step (162) explains the first-order inhibition by H^+ and the small value of the solvent deuterium isotope effect ($k_D/k_H = 1.07$). The salt effect is also in line with the increased dissociation of the weak acid (i.e., the aquo complex) at higher values of ionic strength.

Smidt et al. [177] have explained the inverse dependence of the rate of oxidation of olefin on chloride and hydrogen ion concentrations on the basis of an attack by hydroxide ion on the trihaloethylenepalladate(II) complex, **167**, in the rate-determining step, followed by transfer of hydride from the carbon atom attacked to the neighboring carbon atom, **168**, and the con-certed removal of palladium with its two bonding electrons to form the aldehyde and palladium(0).

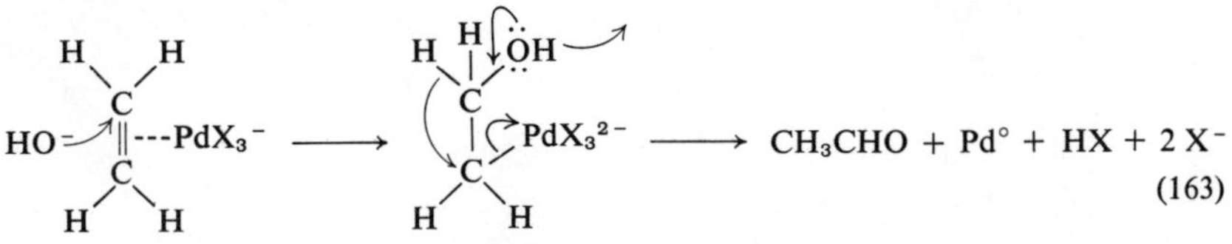

$$\qquad \mathbf{167} \qquad\qquad \mathbf{168}$$

When there are alkyl substituents on the alkene, attack by hydroxide takes place on the more-substituted carbon atom, from which the hydride ion can be readily removed in accordance with reaction (163). Studies of the oxidation of ethylene in D_2O by palladium(II) indicate that the aldehyde formed contains no deuterium. Thus the hydride transfer illustrated in step (163) takes place from the ethylene carbon atom, and the deuterium attached to the attacking deuteroxide anion, OD^-, is dissociated and returns to the solvent.

If the double bond is conjugated with groups such as

$$\text{\textbackslash C=C\textbackslash} \qquad \text{\textbackslash C=O} \qquad -COCOOH \qquad -NO_2$$

the attack by hydroxide ion takes place on the carbon atom which is not bound to these groups [169].

Crotonic acid and its isomer methacrylic acid yield acetone and propion-aldehyde, respectively, on oxidation with palladium(II). These end products are obtained by subsequent decarboxylation of the intermediates **169** and **170**, respectively. In the case of isobutylene, which does not have a hydrogen atom attached to a tertiary carbon atom, the hydride shift in question cannot take place in **171**; the reaction therefore results in the formation of *t*-butanol with no reduction of palladium(II) and as such is similar to the hydration of acetylene (see Chapter 5, Section II).

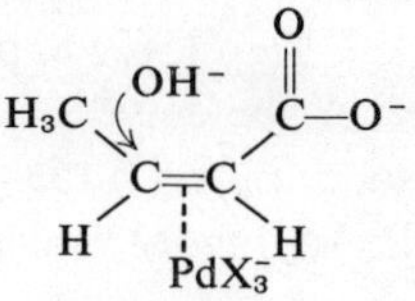

$$\longrightarrow \quad CH_3COCH_3 + Pd(0) + CO_2 + 3\,X^- \quad (164)$$

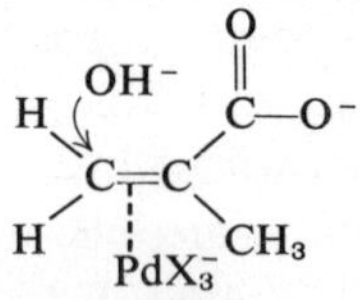

169

$$\longrightarrow \quad CH_3-CH_2-CHO + Pd(0) + CO_2 + 3\,X^- \quad (165)$$

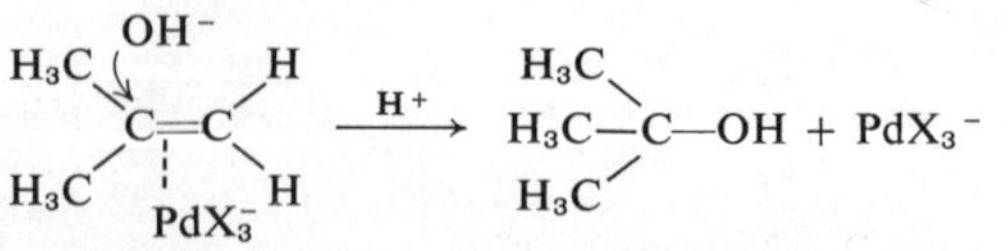

170

$$\xrightarrow{H^+} \quad \begin{array}{c} H_3C \\ H_3C-C-OH + PdX_3^- \\ H_3C \end{array} \quad (166)$$

171

The rate-determining step of the olefin oxidation reaction according to Henry [175] is an oxypalladation reaction, which consists of insertion of the olefin in the Pd—OH bond, as indicated in Eq. (167). Based on the slight

$$\pi\text{-}(C_2H_4)PdCl_2(OH) \longrightarrow \sigma\text{-}HOCH_2CH_2PdCl_2 \quad (167)$$

dependence of the rate of reaction on the olefin structure, it has been proposed [175] that the π–σ rearrangement of the Pd(II) complexes has little carbonium ion character in the transition state. Accordingly, it has been suggested that the π–σ rate-determining oxypalladation step proceeds as a concerted, nonpolar, four-center reaction as indicated by Eq. (168).

$$
\begin{array}{c}
\mathrm{Cl^-} \\
\mathrm{Pd^{2+}} \\
\mathrm{Cl^-}\;\;\;\mathrm{OH^-}
\end{array}
\;
\begin{array}{c}
\mathrm{H} \\
\mathrm{H-C} \\
\mathrm{C-H} \\
\mathrm{H}
\end{array}
\;\xrightarrow[k]{\mathrm{H_2O}}\;
\begin{array}{c}
\mathrm{Cl^-} \;\;\;\mathrm{CH_2-CH_2-OH}\\
\mathrm{Pd^+}\\
\mathrm{Cl^-}\;\;\;\mathrm{OH_2}
\end{array}
\qquad (168)
$$

Aguilo [181] has proposed a mechanism in which the π–σ transition of the palladium(II)–olefin complex takes place by the rotation of the π-bonded alkene about the palladium(II)–alkene bond in such a manner that it moves close to the hydroxide ion bonded to palladium(II). The possibility of such a rotation is supported by the NMR evidence [7], where a large amplitude of rotation of the coordinated alkene about the metal–alkene bond has been established. The antibonding orbitals of the alkene are then exposed to nucleophilic attack by hydroxide ion. Such an attack is possible for the palladium(II) complex because the electron density in the antibonding orbitals of coordinated alkenes is much less than in the platinum(II) complexes [lesser extent of back donation from the pd orbitals of palladium(II) to π^* orbitals of the alkene] (see Chapter 1, Fig. 2). The σ-bonded complex, $[Cl_2PdCH_2CH_2OH]^-$, **172**, seems to have only transitory existence since attempts to prepare ethyl- or propylpalladium complexes of this type with or without stabilizing ligands have been unsuccessful.

The last step of the oxidation of olefin is the fast conversion of the σ-bonded palladium complex **172** to the products. As in Smidt's mechanism [177] a hydride ion shift accompanied by the loss of a hydroxyl hydrogen as a proton must be accounted for in the formation of the aldehyde and palladium-(0). Although many suggestions have been put forward for the dissociation of the σ complex, the most plausible view seems to be the suggestion by Henry [163] that palladium(II) can assist the transfer of hydride from the carbon containing the hydroxyl group to the carbon σ-bonded to palladium(II), as illustrated in reaction sequence (169). That there is no formation of carbonium ions at any stage of the reaction is supported by the absence of glycols in the products of ethylene oxidation. The attractiveness of Henry's mechanism is not only that it is in accord with the general principles of *cis*-olefin insertion as observed in other reactions of olefins such as hydrogenation and the oxo reaction, but also that it assigns an active role to the metal ion in hydride transfer. Various analogous reactions of substituted olefins, presented in

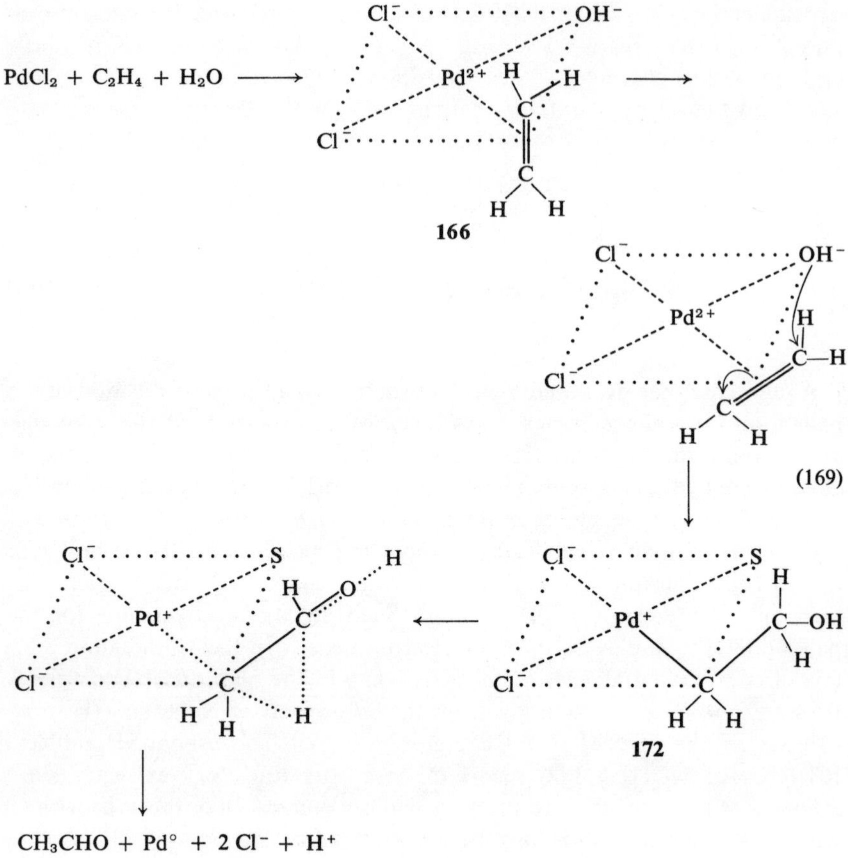

$$PdCl_2 + C_2H_4 + H_2O \longrightarrow$$

Cl⁻·············OH⁻
Pd²⁺ H H
C

C
H   H

166

Cl⁻·············OH⁻
Pd²⁺   H
C—H
C
H   H

(169)

Cl⁻·············S   H
Pd⁺   H   O
C
C
H H   H

Cl⁻·············S   H
Pd⁺   C—OH
C   H
H H

172

$$CH_3CHO + Pd° + 2 Cl^- + H^+$$

reactions (164)–(166), may be explained equally well on the basis of a similar rearrangement, followed by fast hydride shift and decarboxylation.

Okada *et al.* [176] have proposed an equilibrium between σ–π Pd(II) complexes in the oxidation of styrene–palladium(II) in aqueous tetrahydrofuran solution. The rate-determining step of the reaction, unlike Henry's mechanism, is the decomposition of the π complexes **173** and **174** to form the products [Eq. (170)]. Electron-releasing substituents on the phenyl group increase the electron density on the α-carbon atom, facilitate hydride transfer in complex **173**, and hinder hydride transfer in complex **174**, resulting in a higher ratio of ketone to aldehyde. Electron-releasing substituents on the α-carbon atom have the reverse effect. The separate reduction steps indicated by **173a** and **174a** are intended to represent stable intermediates. It is likely that the hydride transfer and metal reduction are concerted reaction steps.

Levanda and Moiseev [174] have investigated the oxidation of ethylene,

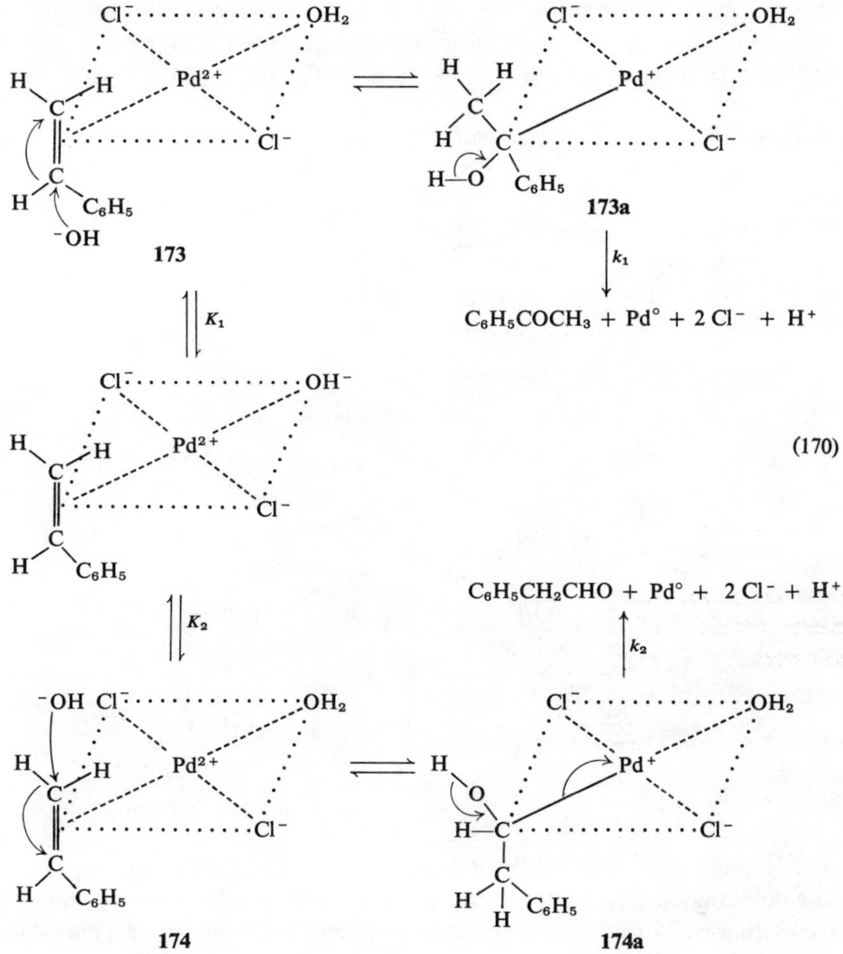

$$C_6H_5COCH_3 + Pd^\circ + 2\,Cl^- + H^+$$

173   **173a**

(170)

$$C_6H_5CH_2CHO + Pd^\circ + 2\,Cl^- + H^+$$

174   **174a**

propylene, and 1-butene by palladium(II) over a wide range of palladium(II) and chloride concentrations and have suggested rate law (171) involving the rate constants k_1 and k_2.

$$\frac{-d[\text{olefin}]}{dt} = k_1 \frac{[\text{PdCl}_4{}^{2-}][\text{olefin}]}{[\text{H}^+][\text{Cl}^-]^2} + k_2 \frac{[\text{PdCl}_4{}^{2-}]^2[\text{olefin}]}{[\text{H}^+][\text{Cl}^-]^3} \quad (171)$$

The rate constant k_1 is the same as that given by Eq. (159) ($k_1 = kK$), and corresponds to the rate constant obtained at lower concentrations ($<0.1\ M$) [166,167] of palladium(II). The mechanism discussed in reactions (160)–(169) applies for the olefin oxidation reactions corresponding to rate constant k_1. At higher concentrations of palladium(II) (0.1–0.2 M), however, a part of the olefin oxidation process proceeds through a second route, corresponding to rate constant k_2. The mechanism that was proposed [174] for the second oxidative pathway is indicated by reactions (160)–(162) and (172)–(174). The

$$C_2H_4 + [PdCl_4]^{2-} \; \underset{}{\overset{K_1}{\rightleftharpoons}} \; [C_2H_4PdCl_3]^- + Cl^- \tag{160}$$
$$\underset{164}{}$$

$$[C_2H_4PdCl_3]^- + H_2O \; \underset{}{\overset{K_2}{\rightleftharpoons}} \; [C_2H_4PdCl_2(H_2O)] + Cl^- \tag{161}$$
$$\underset{165}{}$$

$$[C_2H_4PdCl_2(H_2O)] \; \underset{}{\overset{K_3}{\rightleftharpoons}} \; [C_2H_4PdCl_2(OH)]^- + H_3O^+ \tag{162}$$
$$\underset{166}{}$$

$$[PdCl_4]^{2-} + 166 \; \underset{}{\overset{K_4}{\rightleftharpoons}} \; \begin{bmatrix} Cl^- & Cl^- & Cl^- & OH \\ & Pd^{2+} & & Pd^{2+} & CH_2 \\ Cl^- & Cl^- & & CH_2 \end{bmatrix}^{2-} + Cl^- \tag{172}$$
$$\underset{175}{}$$

$$175 \; \overset{k_2}{\longrightarrow} \; \begin{bmatrix} Cl^- & Cl^- & Cl^- \\ & Pd^{2+} & & Pd^{+} \\ Cl^- & Cl^- & & CH_2CH_2OH \end{bmatrix}^{2-} \tag{173}$$
$$\underset{176}{}$$

$$176 + H_2O \; \overset{fast}{\longrightarrow} \; \begin{bmatrix} Cl^- & Cl^- \\ & Pd^{+} - Pd^{+} \\ Cl^- & Cl^- \end{bmatrix}^{2-} + CH_3CHO + H_3O^+ + Cl^- \tag{174}$$
$$\underset{174}{} \qquad \qquad \longrightarrow \; PdCl_4^{2-} + Pd(0)$$

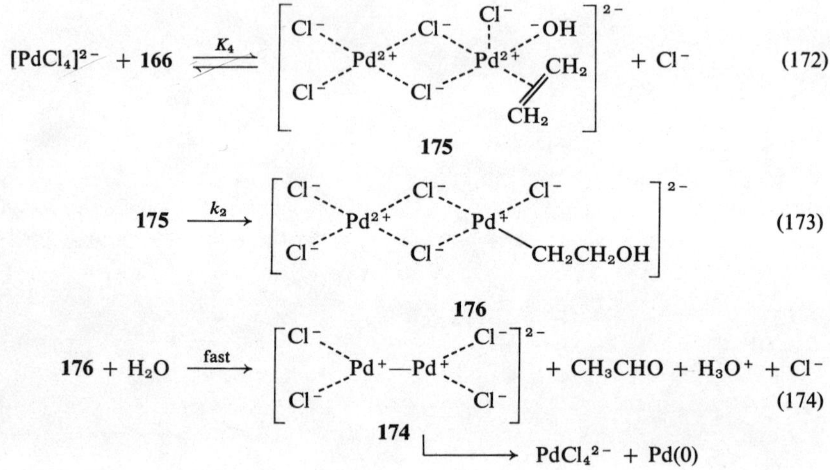

reactive binuclear complex **175** indicated in reaction (172) isomerizes in the rate-determining step (173) to the σ complex **176**. Rapid decomposition of the complex **176** leads to the reaction products indicated in Eq. (174). The activation parameters $\Delta H^{\ddagger}$ and $\Delta S^{\ddagger}$ corresponding to the rate constant k_1 [Eq. (170)] and k_2 [Eq. (173)] have been reported as 19.3 ± 1.6 kcal/mole, −7.1 ± 2.7 eu, 22.4 ± 2.0 kcal/mole, and 6.0 ± 3.0 eu, respectively.

Henry [175] has described oxidation studies of ethylene and propylene in the presence of palladium(II) chloride at a concentration range of 0.01 to 0.2 M and did not obtain a two-term rate expression corresponding to Eq. (171). The reaction kinetics observed agreed with rate expression (159). Henry [175] has attributed the difference in his observation and that of Moiseev [174] to be due to a systematic deviation in the calculation of rate constants from the experimental data in the case of the latter investigation [174]. More thorough and detailed kinetic investigations are apparently needed to answer the questions that have been raised concerning the rate law and mechanism of olefin oxidation in these systems.

A. CATALYTIC SYSTEMS

The oxidation of ethylene by palladium(II) usually results in the formation of palladium metal, as indicated above. On the other hand, the reaction becomes truly catalytic and continuous if palladium(0) is oxidized back to palladium(II). This may be accomplished by adding copper(II) and oxygen to the reaction mixture. The oxidation of ethylene with the palladium(II)–copper(II) catalytic system may be represented [168] by reactions (175)–(177). The palladium(0) formed in reaction (175) is reoxidized to palladium(II) by

$$C_2H_4 + PdCl_2 + H_2O \longrightarrow Pd(0) + 2\,HCl + CH_3CHO \qquad (175)$$

$$Pd(0) + 2\,CuCl_2 \longrightarrow PdCl_2 + 2\,CuCl \qquad (176)$$

$$2\,CuCl + 2\,HCl + \tfrac{1}{2}O_2 \longrightarrow CuCl_2 + H_2O \qquad (177)$$

copper(II) in reaction (176) and in reaction (177) copper(I) is oxidized in turn by molecular oxygen back to copper(II). The oxidation of ethylene is homogeneous when the excess of copper(II) chloride is large enough to rapidly oxidize palladium(0) to palladium(II).

The kinetics and mechanism of the palladium(II)–copper(II)-catalyzed oxidation of ethylene was studied by Matveev et al. [164,165]. The reaction is first order with respect to ethylene, palladium(II), and copper(II) concentrations, inverse first order with respect to chloride and hydrogen ion concentration, and independent of the partial pressure of oxygen. The kinetic expression derived [165] for this system is indicated by Eq. (178).

$$\text{Rate} = \frac{k_2 K_1 K_W [C_2H_4][Pd(II)]}{[Cl^-][H^+]}\left(1 + \frac{k_1}{k_2}K_2[Cu(II)]\right) \qquad (178)$$

This equation corresponds to the mechanistic sequence indicated by reactions (160) and (179)–(181).

$$[PdCl_4]^{2-} + C_2H_4 \xrightarrow{\;K_1\;} [PdCl_3(C_2H_4)]^- + Cl^- \qquad (160)$$
$$\textbf{164}$$

$$[PdCl_3(C_2H_4)]^- + 2\,CuCl_3{}^- \xrightarrow{\;K_2\;} [PdCl_3(C_2H_4)\cdot 2\,CuCl_3]^{3-} \qquad (179)$$
$$\textbf{177}$$

$$[PdCl_3(C_2H_4)\cdot 2\,CuCl_3]^{3-} + OH^- \xrightarrow{\;k_1\;} [PdCl_4]^{2-} + 2\,CuCl_2{}^- + CH_3CHO$$
$$\textbf{177} \hspace{6cm} + H^+ + Cl^- \qquad (180)$$

$$[PdCl_3(C_2H_4)]^- + OH^- \xrightarrow{\;k_2\;} Pd(0) + CH_3CHO + HCl + 2\,Cl^- \qquad (181)$$

Equations (180) and (181) are the rate-determining steps in the presence and absence of copper(II), respectively. In the presence of copper(II), $k_1 \gg k_2$. In the absence of copper(II), k_1 is zero and k_2 becomes rate-determining. Step (160) in the above reaction sequence is similar to the corresponding pre-equilibrium steps in the mechanisms proposed by Smidt [177] and Henry [163] and involves formation of π-ethylene–palladium(II) complex 164. Combination of 164 with $CuCl_3^-$ to form a mixed palladium(II)–copper(II) chloride complex, 177, was suggested, but the probable structure of the complex was not proposed [165]. The rate-determining step [165] was considered to involve attack by hydroxide ion on 177, followed by a concerted intramolecular electron transfer to form the products of the reaction, CH_3CHO, $[PdCl_4]^{2-}$, and $CuCl_2^-$ [reaction (180)]. The copper(I) chloride is believed to be reoxidized in subsequent rapid free-radical reactions with oxygen and hydrogen peroxide as primary and secondary oxidants. The activation energy for the oxidation of ethylene with the palladium(II)–copper(II) catalyst system, measured over the temperature range 30°–90°C, is 13.7 kcal/mole. The activation energy for the oxidation of ethylene by palladium(II) complexes alone [163] is thus reduced by about 6 kcal/mole in the presence of copper(II).

The metal ions of the isoelectronic series consisting of mercury(II), thallium(III), and lead(IV), oxidize olefins to carbonyl derivatives, or to glycols or their esters. In all cases the metal ion undergoes a two-electron reduction, similar to the reduction of palladium(II) to palladium(0) [80,177] and rhodium(III) to rhodium(I) [182]. The oxidation of ethylene by mercury-(II) is heterogeneous, is a relatively complicated reaction, and is not very well understood. The oxidation of 2-hexene by thallium(III), reported by Grinstead [184], leads to the formation of 2-hexanone and 2,3-hexanediol, with reduction of thallium(III) to thallium(I). Since the solvent employed consisted of acetic acid and water, this resulted in partial conversion of the diol to the monoacetate, the yield of the ester decreasing with increasing percentage of water. The overall reaction sequence for the formation of the glycol and the ketone is indicated by Eqs. (182) and (183).

$$RCH{=}CHR + Tl^{3+} + 2\,H_2O \longrightarrow \overset{\displaystyle OH}{\overset{\displaystyle |}{RCH}}{-}\overset{\displaystyle OH}{\overset{\displaystyle |}{CHR}} + Tl^+ + 2\,H^+ \quad (182)$$

$$RCH{=}CHR + Tl^{3+} + H_2O \longrightarrow RCOCH_2R + Tl^+ + 2\,H^+ \quad (183)$$

Grinstead [183] postulated the formation of an olefin–thallium(III) π complex, 178, as indicated in reaction (184). Reaction with hydroxide ion induces a π–σ rearrangement, to give the σ complex 179, from which thallium-(I) dissociates, forming the carbonium ion 180. The latter may react with electron donors to yield various products, as indicated. Also, a proton may be eliminated from the carbon bearing the hydroxide group to give the enol of the corresponding aldehyde or ketone.

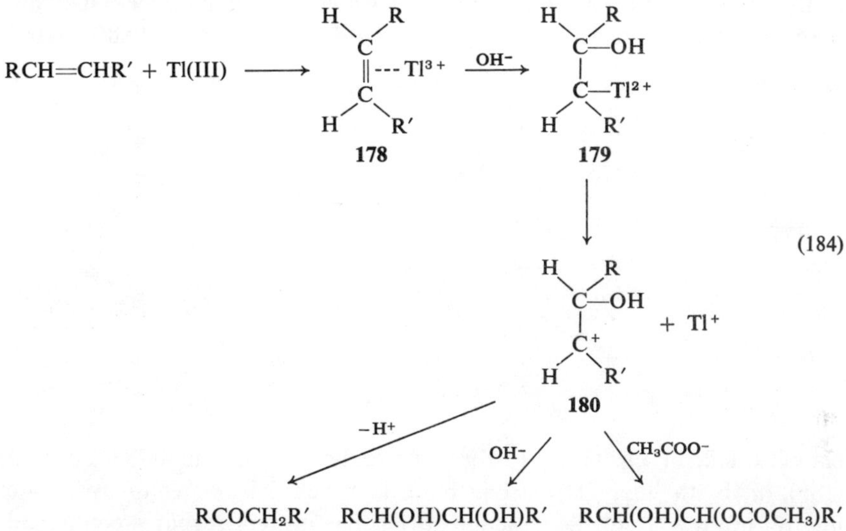

RCH=CHR′ + Tl(III)
H
R
C
C
Tl3+
H
R′
178
OH−
H
R
C—OH
C—Tl2+
H
R′
179
H
R
C—OH
C+
H
R′
+ Tl+
180
−H+
OH−
CH3COO−
RCOCH2R′
RCH(OH)CH(OH)R′
RCH(OH)CH(OCOCH3)R′

(184)

The kinetics and mechanism of oxidation of olefins by thallium(III) has also been studied by Henry [184,185]. The oxidation of ethylene [184], propylene [185], 1-butene [186], *cis*-2-butene [185], and *trans*-2-butene [186] follow the rate law, indicated by Eq. (185).

$$-\frac{d[Tl^{3+}]}{dt} = k[Tl^{3+}][\text{olefin}] \qquad (185)$$

The rate of oxidation of substituted ethylenes was faster than that of ethylene and increased in the order ethylene < *trans*-2-butene < *cis*-2-butene < 1-butene < propylene < isobutene. The molar ratio of carbonyl product to glycol was about 1:1 for ethylene, 3:1 for propylene, and carbonyl compounds were formed exclusively from *cis*- and *trans*-2-butenes. The oxidation of isobutene by thallium(III) produced 35–45% isobutyraldehyde and 55–65% glycol. An olefin–thallium(II) π complex was proposed for the first step of the mechanism as indicated above. This complex is considered to be less stable than the analogous complex formed by palladium(II), because of the higher oxidation state of thallium(III), and consequent less "*d*"-orbital overlap with the olefin. According to the mechanism proposed by Henry, the π complex **178** rearranges to an incipient carbonium ion **181** that is attacked by the solvent to form the species **182**, which is then converted to the σ-bonded metal derivative, **179**, by dissociation of a proton.

The formation of the σ-hydroxyalkylthallium complex **179** from **178** is the proposed rate-determining step of the reaction. The term "oxythallation" has been used to describe this type of reaction [185,186], which may be considered as the insertion of an olefin in the Tl—OH bond. The complex **179** may undergo a hydride shift [similar to that proposed [163] for the palladium(II) σ

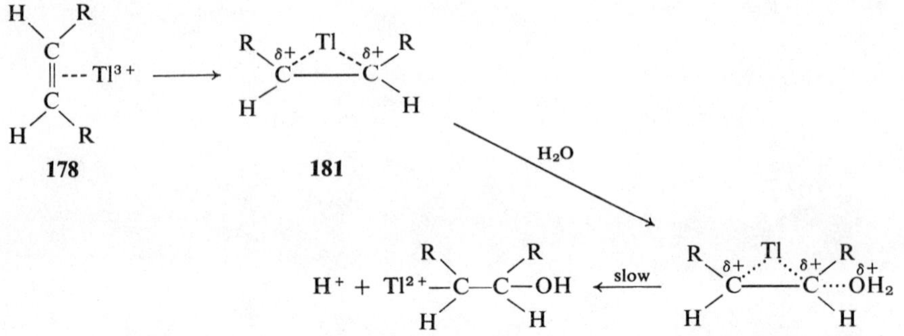

complex **172**, in Eq. (169)] to form the carbonyl compound **183**, reaction (186), or the thallium(III)–carbon bond is attacked by water or hydroxide ion to form the glycol **184** [reaction (187)]. The ratio of carbonyl compound

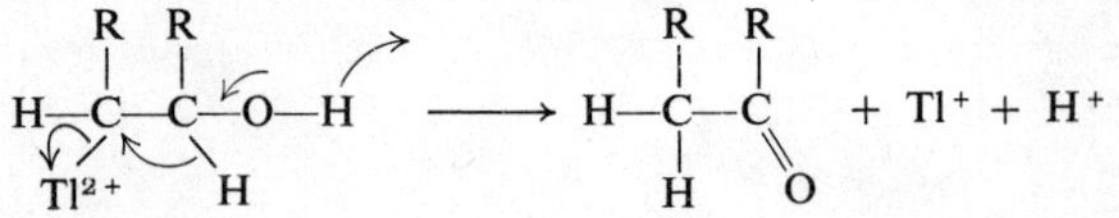

to glycol in the reaction mixture is thus a measure of the extent of hydride shift and hydrolysis reactions that take place. In the case of substituted ethylenes, the hydride shift from the secondary carbon atom is favored to a greater extent than it is in ethylene, resulting in higher ratios of carbonyl compound to glycol.

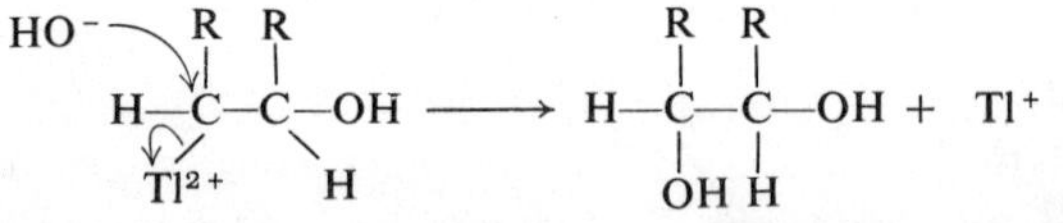

Oxidation by lead(IV) probably proceeds through a similar mechanism, involving initial complex formation with olefin. The lead(IV) complex is probably very much less stable than that of thallium(III); in fact, because of their low stabilities, lead(IV)–olefin complexes are not well known. The mechanism of formation of a diacetate or glycol monoester in the reaction of isobutylene with lead tetraacetate may be rationalized with the following reaction sequence [187]:

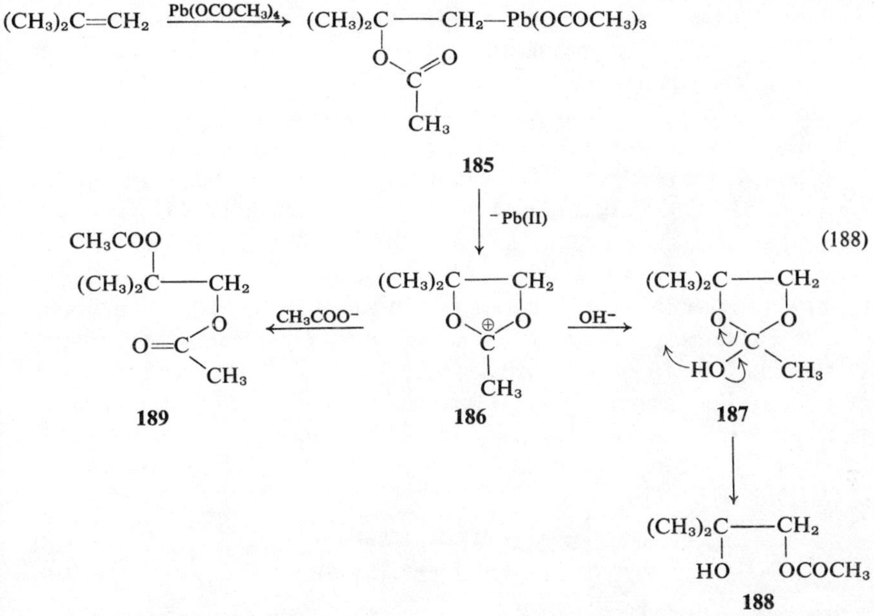

$(CH_3)_2C{=}CH_2 \xrightarrow{Pb(OCOCH_3)_4} (CH_3)_2C{-}{-}{-}CH_2{-}Pb(OCOCH_3)_3$

185

(188)

189   **186**   **187**

$(CH_3)_2C{-}{-}{-}CH_2$
$\quad\quad HO \quad\quad OCOCH_3$

188

In reaction (188) lead(IV) is reduced to lead(II) and the resulting carbonium ion, **186**, reacts either with water to form the hemiester **188** or with an acetate anion derived from another molecule of $Pb(OCOCH_3)_4$ to give the diacetate **189**.

II. Hydration of Alkynes

Acetylene is more acidic than ethylene; consequently it shows higher reactivity toward nucleophiles and less affinity for electrophiles. Various protonic acids that are specific catalysts for the hydration of olefins have very low reactivity in the conversion of acetylene and its derivatives to hydrated compounds. Nucleophiles, on the other hand, add readily to alkynes, the relative reactivities of the nucleophiles decreasing in the order iodide > bromide > chloride > cyanide > hydroxide > acetate. Addition of an iodide to the triple bond requires no catalyst; the other nucleophiles, however, require metal ion catalysis for the reaction to proceed with any reasonable speed.

Metal ions that are effective catalysts for the hydration of acetylene and other alkynes are mercury(II), copper(I), silver(I), ruthenium(III), and rhodium(III). The type of metal ion catalyst required for nucleophilic addition to alkynes depends on the extent of interaction between the nucleophilic reagent and the metal ion, and between the alkyne and the metal ion. When the metal ion has too strong an affinity for the nucleophile, the latter will not add to the alkyne. If the metal ion undergoes strong interaction [187] with the alkyne, the reaction results in the dissociation of a proton and the formation of acetylides or stable organometallic compounds.

Hydration of an alkyne takes place by the initial activation of the —C≡C— triple bond as the result of the formation of a π complex in the rate-determining step [158,188,189]. Insertion of the alkyne molecule in the M—OH bond then takes place in a manner similar to the *cis*-insertion mechanisms of alkenes in hydrogenation or oxo reactions. Dissociation of the metal ion in the final step results in the formation of a hydrated alkyne. The formation of soluble π complexes of copper(I) and mercury(II) with acetylene has been substantiated by the potentiometric work of Temkin *et al.* [190] and Vestin *et al.* [191]. Equilibrium constants for reactions (189) and (190) in aqueous solution have been reported as 7.36×10^5 [copper(I)] and 7.75×10^5 [mercury(II)] at 25°C in acid solution by Vestin *et al.* [191] and by Temkin *et al.* [190], respectively.

$$Cu^+ + C_2H_2 \rightleftharpoons [C_2H_2 \cdot Cu]^+ \tag{189}$$

$$Hg^{2+} + C_2H_2 \rightleftharpoons [C_2H_2 \cdot Hg]^{2+} \tag{190}$$

The reaction of $CuCl_2{}^-$ with acetylene [188] results in the formation of vinylacetylene, acetaldehyde, and vinyl chloride. Partial reduction of copper(I) to copper(0) also takes place during the reaction. In very acidic solutions, however, copper acetylide is formed. The acetylide precipitates from solution and thus complicates the hydration of acetylene with copper(I) as catalyst, a process that would otherwise take place homogeneously.

Addition of acetylene to mercury(II) sulfate solution in sulfuric acid solution or to mercuric chloride in hydrochloric acid solution, results in the formation of a series of products, including the acetylide, HgC_2; organometallic compounds, $C_2H_2HgCl_2 \cdot HgCl$; chlorovinylmercury; aldehydes and ketones; and mercuration products of aldehydes and ketones. Relative amounts of products obtained vary with hydrogen ion concentration, concentration of substrate, and concentration of mercury(II) sulfate. Partial reduction of mercury(II) to mercury(0) also takes place.

Temkin *et al.* [190] studied the drop in the potential of the mercury electrode when acetylene is added to a solution of mercury(II) sulfate in sulfuric acid and resolved the observed effect into reversible and irreversible steps. The reversible potential drop of the mercury electrode was postulated to be due to the formation of a series of complexes according to Eqs. (191)–(193). In the presence of 1.5–3.0 M sulfuric acid, the complex $[HgC_2H_2]^{2+}$

(190) is the main species in solution. These are the same conditions under which the maximum catalytic activity of mercury(II) was observed at 90°C.

$$Hg^{2+} + C_2H_2 \xrightleftharpoons{K_1} HgC_2H_2^{2+} \tag{191}$$
$$190$$

$$HgSO_4 + C_2H_2 \xrightleftharpoons{K_2} HgSO_4C_2H_2 \tag{192}$$

$$Hg(SO_4)_2^{2-} + C_2H_2 \xrightleftharpoons{K_3} Hg(SO_4)_2C_2H_2^{2-} \tag{193}$$

Hydration of acetylene was therefore postulated as occurring through hydration of the mercury(II)–acetylene complex, 190, as indicated in Eq. (194).

$$[HgC_2H_2]^{2+} \xrightarrow[k]{H_2O} CH_3CHO + Hg^{2+} \tag{194}$$

This rate expression applies to the concentration range 2.5×10^{-2} to 1×10^{-1} M mercury(II) sulfate in 10% sulfuric acid at 90°C. At higher concentrations the reaction becomes heterogeneous because of the precipitation of a solid phase.

Budde and Dessy [192] studied the hydration of phenylacetylene with mercury(II) perchlorate as catalyst in a water–dioxane–perchloric acid medium. When the ratio of alkyne to mercury(II) ion was 1:1, a 50% yield of acetophenone was obtained. Increasing the ratio of substrate to mercury(II) increased the yield of acetophenone, while decreasing the ratio led to lower conversion of alkyne to the hydration product. When mercury(II) is mixed with phenylacetylene a yellow color is observed, involving an absorption band at 315–330 mμ. As the concentration of acetophenone builds up in solution, there is a gradual fading of the yellow color and the solution finally becomes colorless. The formation of this yellow color is attributed to the formation of a 2:1 complex of mercury(II) with phenylacetylene in the reaction mixture. The formation of the yellow complex was also noticed by Lemaire and Lucas [193] in the mercury(II) ion-catalyzed addition of acetic acid to 3-hexyne. The formation of a 2:1 alkyne–mercury(II) complex is further substantiated [192] by the reaction of diphenylacetylene with mercury-(II). The stoichiometry of the complex was established [192] by measuring the absorption at 320 mμ against the molar ratio of diphenylacetylene to mercury-(II). A break in the plot of absorption versus molar ratio at [alkyne]/[Hg(II)] = 2 was attributed to the formation of a 2:1 complex. As the amount of water increases in the reaction mixture (to more than 3.1 mole per mole dioxane), a heterogeneous phase appears. The 2:1 complex, 191, of phenylacetylene with mercury(II) is similar to the structure suggested by Budde and Dessy [192].

In addition to copper(I) and mercury(II), palladium(II) and ruthenium(III)

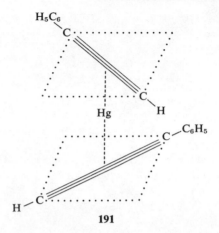

191

may also be used as catalysts for the hydration of alkynes. The use of palladium(II) chloride with methyl acetylene produces a complicated heterogeneous mixture of palladium(0) crotonaldehyde, acetaldehyde, and polymers of crotonaldehyde [194].

A detailed study of ruthenium(III) catalysis in the hydration of alkynes was conducted by Halpern *et al.* [189]. Acetylene is taken up by a 0.1 M solution of ruthenium(III) in 5 M hydrochloric acid at 50°C to give mainly acetaldehyde, along with a small amount of crotonaldehyde. Methylacetylene and ethylacetylene react under the same conditions to give acetone and methyl ethyl ketone, respectively. In the case of acetylene, the rate law given in Eq. (195) was observed.

$$-\frac{d[C_2H_2]}{dt} = k[C_2H_2][Ru(III)] \tag{195}$$

$(k = 2 \times 10^{-3}$ mole liter^{-1} min^{-1} at 50°C)

The rate constant k increases with hydrochloric acid concentration up to 4 M and then falls off at higher concentrations. In a 5 M hydrochloric acid solution, the active ruthenium(III) species present in solution are [Ru(H$_2$O)$_2$-Cl$_4$]$^-$ and [Ru(H$_2$O)Cl$_5$]$^{2-}$. The decline in rate at high chloride ion concentration is considered due to the replacement of all water molecules in the coordination sphere of ruthenium(III) by chloride ions, resulting in the formation of the relatively inactive [RuCl$_6$]$^{3-}$ complex. It was thus established that hydration of acetylene requires at least one water molecule coordinated to ruthenium(III). Reaction sequence (196) was put forward to explain these observations [189]. The rate-determining step in this mechanism is the formation of a π complex, **192**, between ruthenium(III) and the alkyne. An alkyne insertion reaction then takes place in a subsequent fast step, accompanied by

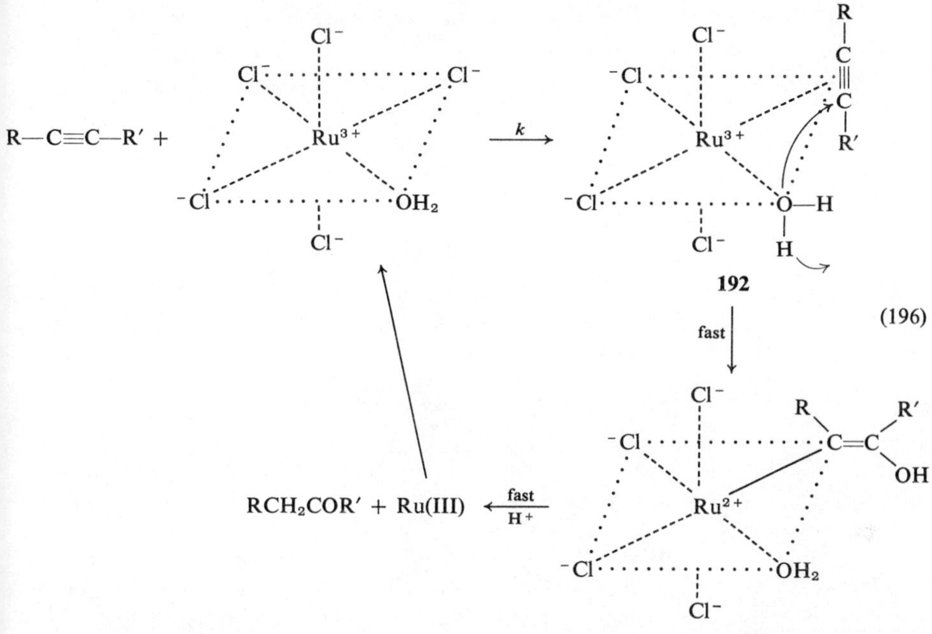

$$R-C\equiv C-R' +$$

192

193

(196)

$$RCH_2COR' + Ru(III) \xrightleftharpoons[H^+]{fast}$$

the liberation of a proton. The σ complex then dissociates in a fast step, resulting in the formation of aldehyde or ketone and regeneration of the catalytic metal ion species.

James and Remple [195] have proposed a similar mechanism for the hydration of alkynes catalyzed by rhodium(III) chloride. It may be readily seen that unlike the oxidation of alkenes to aldehydes or ketones, hydration of alkynes involves no change in the oxidation state of the metal ion. The net reaction involves addition of hydrogen and hydroxyl ions across the triple bond, followed by rearrangement of the bonds to form an aldehyde or ketone. The decline in the rate of hydration at higher chloride ion concentration is due to the formation of neutral and anionic species of ruthenium(III) that are substitution-inert.

III. Alkene or Alkyne Insertion into Metal–Carbon Bonds

Alkene insertion into metal–carbon bonds is a general palladium(II)-catalyzed reaction in which an alkyl or aryl group originally bonded to palladium migrates to one unsaturated carbon atom of a π-bonded alkene, which in turn becomes σ-bonded to the metal through the adjacent carbon atom [196]. The palladium(II) σ-bonded complex **194** can undergo either an elimination reaction (197a) to form a vinyl derivative **195** or a metal displacement reaction (197b) may take place to form a saturated product, **196** [197]. In reaction (197b) palladium(II) accepts two electrons and is converted into

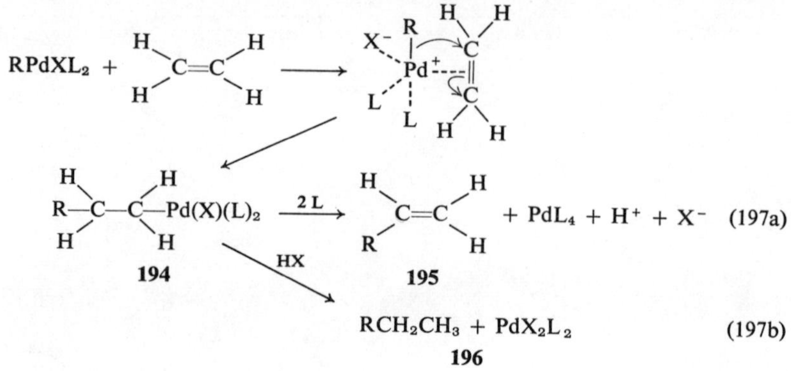

$$RCH_2CH_3 + PdX_2L_2 \qquad (197b)$$
196

(L = PPh$_3$ R = C$_6$H$_5$, alkyl)

palladium(0) (i.e., PdL$_4$), which is stabilized by combination with triphenylphosphine. In the absence of triphenylphosphine metallic palladium is precipitated. The palladium(0) complex PdL$_4$ can be converted into RPdX(L)$_2$ by oxidative addition of RX. Thus the overall reaction can be made catalytic, with RX and CH$_2$=CH$_2$ being converted to RCH=CH$_2$ and HX.

In reaction (197a) the alkyl or aryl group R may be replaced by hydroxide,

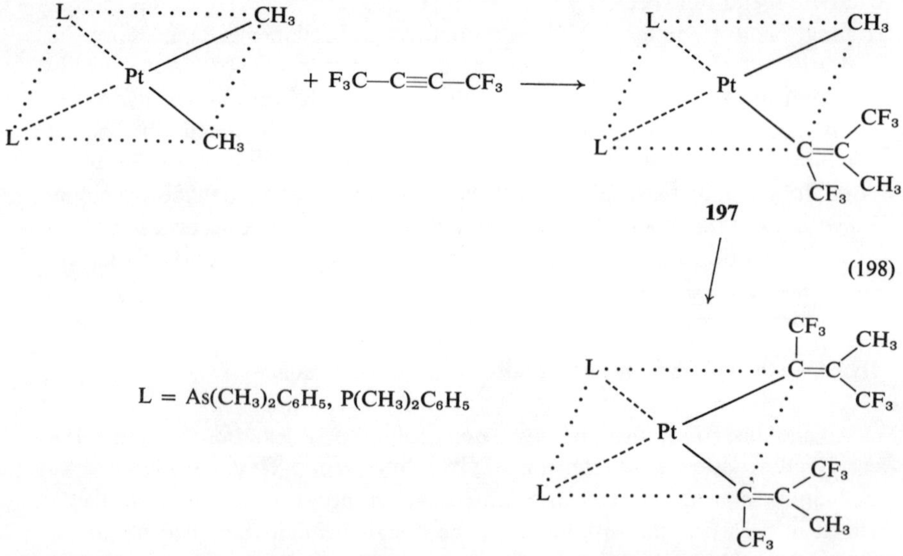

L = As(CH$_3$)$_2$C$_6$H$_5$, P(CH$_3$)$_2$C$_6$H$_5$

197

(198)

198

alkoxide, acetate, amine, and carbanion residues to give a variety of substituted vinyl compounds.

An interesting insertion reaction of alkenes and alkynes into the platinum–carbon bond has been reported by Clark and Puddephant [198]. cis-$Pt(CH_3)_2$-L_2 reacts with hexafluorobut-2-yne at room temperature to give σ-bonded alkene complexes **197** and **198** resulting from the insertion of the alkyne into one or both methyl–platinum bonds.

The reaction of tetrafluoroethylene with cis-$Pt(CH_3)_2L_2$ required a higher temperature and gave products of insertion of C_2F_4 into one or both methyl–platinum bonds.

Alkene and alkyne insertion into metal–carbon bonds provides a convenient method of carbon–carbon bond formation in the synthesis of hydrocarbon derivatives [199,199a]. Many such reactions have been reported by Maitlis [200,200a], to whose book the reader is referred for further information.

6

Multiple Insertion Reactions of Alkenes and Alkynes

I. Introduction

Chapters 2–5 have dealt with reactions in which alkenes and alkynes are inserted in a metal–ligand bond by activation of the hydrocarbon through metal π-complex formation. Multiple insertion reactions consist of a sequence of steps in which a series of alkene or alkyne molecules are inserted in a metal–ligand bond, leading either to long polymer chains of high molecular weight or to closed-ring systems. In some cases the chain may stop at an integral number of monomer units, leading to low molecular weight oligomers and co-oligomers. In most cases the nature of the metal complex catalyst and the ligands bound to it control the molecular weights and stereoisomerism of the polymers or oligomers formed.

The discovery by Ziegler and Natta and co-workers [201–206] of the formation of stereospecific polymers in multiple insertion reactions of alkenes is one of the major developments in metal ion-catalyzed polymerization reactions. Catalysis of these polymerization reactions by transition metal complex compounds in conjunction with aluminum alkyls or aluminum alkyl halides is referred to as "Ziegler" or "Ziegler-Natta" catalysis. The transition metal ion originally used by Ziegler *et al.* [201] was titanium(IV); the list now includes soluble and insoluble complexes of titanium(IV), vanadium(IV), vanadium(III), zirconium(IV), chromium(III), iron(II), iron(III), cobalt(II) cobalt(III), and nickel(II). Metal alkyls other than aluminum, such as zinc diethyl and lithium alkyls, are also used in Ziegler catalyst systems. Thus catalysts with various degrees of activity and specificity for polymerization or copolymerization of alkenes are obtained by suitable combination of a tran-

99

sition metal complexes with aluminum alkyls, and/or aluminum alkyl halides. Polymers with various degrees of crystallinity, toughness, and impact strength have been obtained from these Ziegler catalyst-controlled polymerizations.

Natta [202,203,206] was the first to use the terms "isotactic" and "atactic" for stereoregular polypropylenes. Stereoregular polymers of butadiene and isoprene have been of considerable value to the rubber industry. Synthetic polyisoprene or polybutadiene with 99% *cis*- or *trans*-1,4- or 1,2-structure can be obtained by the use of the new catalytic systems listed above. Sections II–IV treat various aspects of polymerization of alkenes, dienes, and alkynes by Ziegler catalysis.

Besides the formation of long polymer chains, suitable metal catalysts such as those of ruthenium(III) [207,208], rhodium(III) [209], palladium(II) [55,210], cobalt(II) [211], and nickel(II) [212] also produce oligomers of ethylene and other alkenes and alkadienes. The co-oligomerization of olefins with dienes by the use of rhodium(I) [213,214] and cobalt(0) [215] complexes as catalysts has also been reported. The formation of oligomers or co-oligomers permits the synthesis of polyenes of desired chain lengths and configurations from alkenes and dienes. Pertinent aspects of linear oligomerization and co-oligomerization of alkenes and dienes are described in some detail in Sections IV and V. Oligomerization leading to cyclization and aromatization will be the subject matter of Sections VI and VII.

II. Polymerization of Ethylene and α-Olefins

A. ZIEGLER CATALYSIS

In 1955 Karl Ziegler and co-workers [201] announced the synthesis of polyethylene by polymerization of ethylene in a saturated hydrocarbon solvent through the catalytic effect of aluminum triethyl and titanium tetrachloride at 50°–150°C (50–500 lb/inch²). The polyethylene prepared with this Ziegler catalyst had physical properties completely different from the conventional high-pressure product. It has a density [216] of 0.96 g/cm³, stiffness of 140,000 lb/inch², tensile strength of 400 lb/inch², and contained less than two branches per 1000 carbon atoms. The high-pressure polyethylene known previously has a density of 0.92 g/cm³, stiffness of 25,000 lb/inch², tensile strength of 200 lb/inch², and about 20 to 30 branches per 1000 carbon atoms. Another significant advance in the preparation of polymers with Ziegler catalysts was the preparation of stereospecific crystalline polypropylene by Natta [203,206] from titanium trichloride and aluminum alkyl in a hydrocarbon medium. The catalytic titanium and aluminum alkyl system is, however, heterogeneous, and the stereospecificity of the polypropylene product is due mainly to the presence of a layer lattice of the catalyst that determines the configuration of the growing polymer chain. In the case of polypropylene, three types of steric arrangements are thus possible:

1. Same steric arrangement of monomer units by a head-to-tail chain arrangement, referred to as *isotactic*

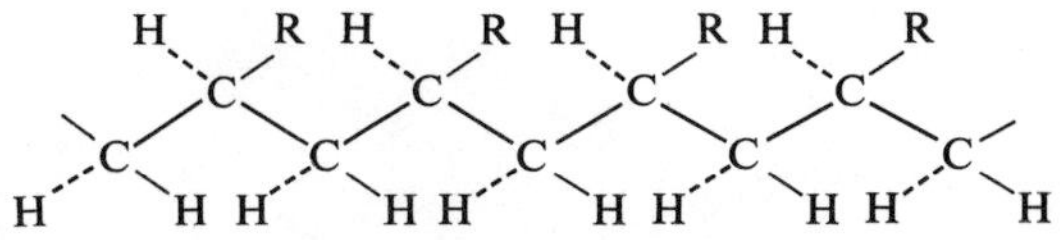

2. Alternately opposite steric arrangement of monomers, referred to as *syndiotactic*

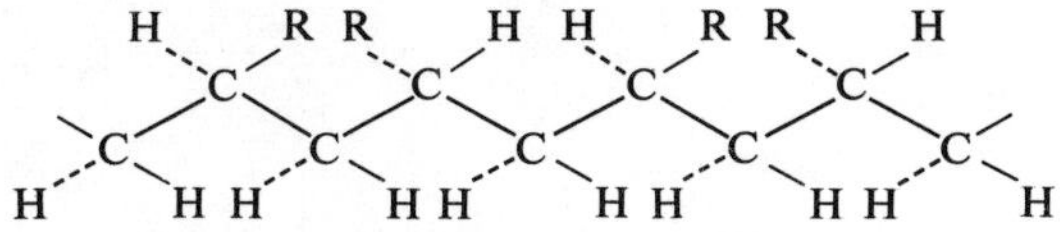

3. *Atactic*, or at-random arrangement of monomer units

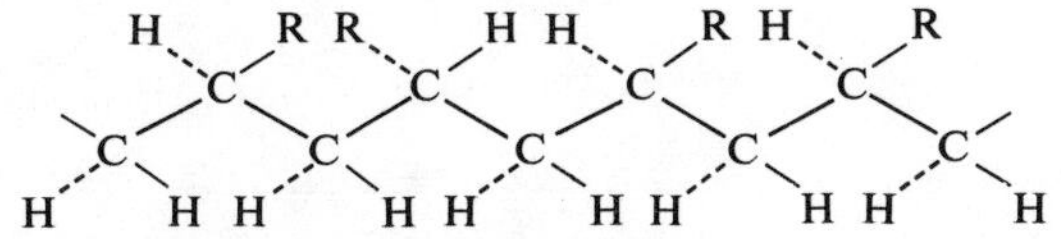

The steric purity of the polymer in each case is controlled by the physical state of the catalyst system, and by the temperature, pressure, and solvent employed. Stereospecific polymers of α-olefins have been obtained [204] only with heterogeneous catalysts composed mainly of $TiCl_3$ and AlR_3 (R = alkyl). Homogeneous hydrocarbon-soluble Ziegler catalysts furnish only nonstereospecific polymers. A list of catalysts for the polymerization of alkenes is presented in Table III.

Entries 1–3 in Table III represent the heterogeneous catalysts most suitable for stereospecific polymerization of the alkenes. Entries 4–16 involve soluble Ziegler catalysts for the polymerization of a number of alkenes and cyclobutene. With the exception of the low temperature ($-78°C$) polymerization of propylene [217] that furnishes syndiotactic polypropylene (entry 13, Table III), polymerization in most of the homogeneous cases yields only nonstereospecific polymers. Non-Ziegler-type polymerization catalysts that contain nickel(0) [218], palladium(II) [219,220], silver(I) [221], rhenium(V) [222], and rhodium(III) [223] usually catalyze stereospecific polymerization of alkenes. Ruthenium(III) [224] and silver(I) [221] form water and alcohol-soluble polymerization catalysts that have the added advantage of being effective at ordinary temperature and pressure.

TABLE III

CATALYSTS FOR THE POLYMERIZATION OF ALKENES BY MULTIPLE INSERTION REACTIONS

No.	Catalyst	Type of medium	Substrate	Temp. (°C)	Products	Ref.
1	$TiCl_3$ + AlR_3	Heterogeneous	Propylene	60–90	Isotactic polypropylene	221
						204
2	$TiCl_3$ + AlR_3	Heterogeneous	2-Butene (*cis* or *trans*)		High mol. wt. polymer with recurring 2-butene units	228
3	$TiCl_3$ + AlR_3	Heterogeneous	Olefins and haloolefins	50–70	Copolymers	229
4	(*sec*-Bu)$_4$Ti + AlR_3	Homogeneous (hydrocarbon)	Ethylene	20–50	Low mol. wt. polyethylene	234
5	$(C_5H_5)_2TiCl_2$ + AlR_3	Homogeneous (hydrocarbon)	Ethylene	30	Polyethylene	234
6	$(C_5H_5)_2TiCl_2$ + AlR_3	Homogeneous (hydrocarbon)	*d*-Styrene		Polystyrene	231
						232
7	$(C_5H_5)_2TiCl_2$ + $Al(CH_3)_2Cl$	Homogeneous (hydrocarbon)	Ethylene	30	Polyethylene	226
8	(*n*-Butoxy)$_4$Ti + $Al(C_2H_5)_2Cl$	Homogeneous (hydrocarbon)	Vinyl chloride	30	Vinyl polymer	233
9	VCl_4 + Al(hexyl)$_3$	Homogeneous (hydrocarbon)	Ethylene + butene	−30	Copolymer	240
10	VCl_4 + Al(phenyl)$_3$	Homogeneous (hydrocarbon)	Ethylene	60	Polyethylene	236
11	VCl_4 + Sn(phenyl)$_4$ + AlX_3 (X = Cl Br,I)	Homogeneous (hydrocarbon)	Ethylene	60	Polyethylene	237
						241
12	Tris(2,4-pentanediono)–V(III) + $Al(C_2H_5)_2Cl$	Homogeneous (hydrocarbon)	Ethylene + 2-butene	−30	Copolymers	240

	Catalyst	System	Monomer	Temp.	Polymer	Ref.
13	Tris(2,4-pentanediono)-V(III) + Al(C$_2$H$_5$)$_2$Cl	Homogeneous (hydrocarbon)	Propylene	−78	Syndiotactic polypropylene	217
14	Bis[2,4-pentanediono]-oxovanadium(IV) + Al(C$_2$H$_5$)$_2$Cl	Homogeneous (hydrocarbon)	Ethylene	20	Polyethylene	238
15	Tris(2,4-pentanediono)-Cr(III) + Al(C$_2$H$_5$)$_2$Cl	Homogeneous (hydrocarbon)	Cyclobutene	−30	Cyclobutylenomers	238
16	Tris(2,4-pentanediono)-Cr(III) + Al(C$_2$H$_5$)$_2$Cl	Homogeneous (hydrocarbon)	Ethylene	20	Polyethylene	239
17	Ru(III) chloride	Homogeneous (ethanol)	3-Methylcyclobutene	20	Polybutadiene 1,4-*trans*, 85% 1,4-*cis*, 15%	224
18	Ru(III) chloride	Homogeneous (water)	3-Methylcyclobutene	50	Polybutadiene 1,4-*trans*, 35% 1,4-*cis*, 65%	224
19	Ru(III) chloride	Homogeneous (water)	Cyclobutene	50	Polybutadiene 1,4-*trans*, 50% 1,4-*cis*, 50%	224
20	Ru(III) chloride	Homogeneous (ethanol)	Cyclobutene	20	Polybutadiene 1,4-*trans*	224
21	Rh(III) chloride + sodium dodecyl benzoate	Homogeneous (water)	Cyclobutene	50	Polycyclobutylenomers erythro diisotactic	223
22	Bis(allyl)NiX$_2$	Homogeneous (ethanol)	Cyclobutene	50	Polycyclobutylenomers erythro diisotactic	218
23	AgClO$_4$	Homogeneous (ethanol)	Vinyl chloride	50	Vinyl polymer	221
24	ReCl$_5$	Homogeneous (ethanol)	Styrene	—	Polystyrene	222

Insight into the nature and catalytic activity of Ziegler catalysts is provided by the study of hydrocarbon-soluble systems amenable to quantitative measurements. One such example is the catalyst obtained [225–227] by the interaction of bis(cyclopentadienyl)titanium dichloride with alkyl aluminum compounds. The catalyst is active only in the polymerization of ethylene, and its activity is comparable to the activities of insoluble Ziegler catalysts [204,221,228,229] used for the preparation of linear high molecular weight polyethylene with a narrow molecular weight distribution. On the basis of spectral evidence [230] the existence of three complexes, $(C_5H_5)_2TiCl_2(C_2H_5)_2$-AlCl, $(C_5H_5)_2(C_2H_5)TiCl(C_2H_5)AlCl_2$, and $(C_5H_5)_2TiCl(C_2H_5)AlCl_2$, has been found in solutions containing bis(cyclopentadienyl)titanium dichloride and an aluminum ethyl. The titanium complexes referred to are probably formed in solution by reactions of the type illustrated by Eqs. (199)–(201).

$$2\ (C_5H_5)_2TiCl_2 + [(C_2H_5)_2AlCl]_2 \longrightarrow 2\ (C_5H_5)_2TiCl_2 \cdot (C_2H_5)_2AlCl \qquad (199)$$
$$\mathbf{199}$$

$$(C_5H_5)_2TiCl_2 \cdot (C_2H_5)_2AlCl \longrightarrow (C_5H_5)_2(C_2H_5)TiCl \cdot (C_2H_5)AlCl_2 \qquad (200)$$
$$\mathbf{200}$$

$$(C_5H_5)_2Ti(C_2H_5)Cl \cdot (C_2H_5)AlCl_2 \longrightarrow (C_5H_5)_2TiCl \cdot (C_2H_5)AlCl_2 + C_2H_5 \cdot (201)$$
$$\mathbf{201}$$

When a solution of bis(cyclopentadienyl)titanium dichloride in benzene is added to alkylaluminum chloride, the sandwich dichloride disappears very rapidly and a peak at 515 mμ due to species **199** appears. Species **199** gradually transforms in solution to species **200** (absorption maximum at 500–530 mμ). Species **200** then decomposes to the tervalent titanium complex **201**. The decomposition of **200** to **201** may be followed conveniently in solution either at 520 or 750 mμ. Thus the initial rate of polymerization of ethylene may be correlated either with the rate of formation of **200** at 0°C, or its decomposition at 30°C to **201**. It was postulated [230a] that species **200** is the catalytically active form that converts ethylene to high molecular weight polymers in solution.

Overberger *et al.* [231,232] have postulated the equilibria (202) and (203) to account for the catalytic activity of bis(cyclopentadienyl)titanium dichloride and diethylaluminum chloride in benzene solution in the polymerization of deuterostyrene, *p*-methylstyrene, and *m*-methylstyrene. The ion pair **202**

$$(C_5H_5)_2TiCl_2 \cdot Al(C_2H_5)_2Cl \xrightleftharpoons{\ C_6H_6\ } (C_5H_5)_2TiCl^+ \dots Al(C_2H_5)_2Cl_2{}^- \qquad (202)$$
$$\mathbf{202}$$

$$(C_5H_5)_2TiCl^+ \dots Al(C_2H_5)_2Cl_2{}^- \rightleftharpoons (C_5H_5)_2Ti(C_2H_5)Cl \cdot Al(C_2H_5)Cl_2 \qquad (203)$$
$$\mathbf{203}$$

formed in reaction (202) is converted in a unimolecular step (203) to an active species **203** by the alkylation of titanium. In the polymerization of deutero-styrene, a deuterium atom probably replaces the halogen in **203**, the resulting deuterium–proton transfer steps thus accounting for the large deuterium isotope effect observed for this reaction.

Yamazaki *et al.* [233] employed a soluble catalyst obtained from tetra-*n*-butoxytitanium(IV) and diethylaluminum chloride for the polymerization of vinyl chloride at 30°C. The catalytic activities of aluminum alkyls in combination with Ti(*n*-butoxide)$_4$ decrease in the order, $AlCl_2(C_2H_5)$ > AlCl-$(C_2H_5)_2$ $\gg$ $Al(C_2H_5)_2(OC_2H_5)$ > $Al(C_2H_5)_3$ > $AlCl_3$. Bawn and Symcox [234] and Bawn and Ledwith [235] used 2-butyl titanate and aluminum triethyl for the polymerization of ethylene to low molecular weight polymers. (The molecular weight obtained at 20°–50°C and 0–2 atm was in the range of 400–500.) The reaction is inhibited by alkyl halides, alcohols, and other polar solvents.

Combination of triphenylaluminum or triisobutylaluminum with vanadium tetrachloride forms soluble catalysts that polymerize ethylene in cyclohexane solution at 60°C and 1 atm pressure to linear high molecular weight polyethylene with less than one (terminal) methyl group per 1000 carbon atoms [235]. Addition of aluminum alkyls to vanadium tetrachloride, in a molar ratio of 3:1, in cyclohexane solution causes complete reduction of vanadium(IV) to vanadium(III) species in less than a minute [236,237]. The vanadium species present in the pink solution so obtained has been shown by polarographic analysis to be exclusively divalent. On the other hand, vanadium dichloride or dibromide are inactive as catalysts in the absence of an aluminum alkyl. The addition of an aluminum halide or aluminum alkyl in cyclohexane causes rapid dissolution of vanadium(II) and the formation of an active catalyst that has been assigned structure **204** [236,237]. Complex

$$(R)X^- \quad X^-$$
$$Al^{3+(+)} \quad V^+\!\!-\!\!R$$
$$(R)X \quad X^-$$

204

(X = halogen; R = alkyl or aryl)

204 is a donor–acceptor type complex and is readily destroyed in solution by the addition of a stronger base, such as an ether. The function of the aluminum alkyl seems to be the reduction of vanadium to the active divalent state, with subsequent alkylation of the vanadium(II) ion.

Alkylation of vanadium in the active catalytic species is analogous to the reaction of diphenylmercury with vanadium oxytrichloride or vanadium tetrachloride in cyclohexane solution [238]. Interaction of the vanadium halide and diphenylmercury leads to an intermediate arylvanadium halide, which then rapidly decomposes to diphenyl and a lower valence vanadium compound according to reaction (204).

$$C_6H_5VCl_n \longrightarrow \tfrac{1}{2}C_6H_5\text{—}C_6H_5 + VCl_{n-x}^{+x} + x\,Cl^- \qquad (204)$$

A mixture of aluminum halide, vanadium halide, and tetraphenyltin in cyclohexane forms a clear solution [237] that produces low-pressure polymerization of ethylene even at concentrations of vanadium as low as one part in 10^6 parts of the substrate [230]. The polyethylene so formed is linear, possesses a high molecular weight (65,000–120,000) with a narrow range of molecular weight distribution, and contains less than one unsaturated group for every 10,000 carbon atoms. This catalyst is, however, ineffective for the polymerization of propylene. A narrow molecular weight distribution of the polymer suggests that the transition metal ion present in the system may act as a polymerization catalyst by the formation of coordinated olefin complex intermediates. (Wider distributions of molecular weights are usually obtained in free-radical or Friedel-Crafts type polymerization mechanisms.)

Olive and Olive [239] have used bis(2,4-pentanediono)oxovanadium(IV) and tris(2,4-pentanediono)chromium(III) as transition metal components, together with $(C_2H_5)_2AlCl$ in hydrocarbon solvents to form a catalyst system for the polymerization of ethylene. From ESR and magnetic susceptibility measurements it was concluded [239] that the active catalytic species are the vanadium(II) and chromium(II) complexes **205** and **206**, respectively.

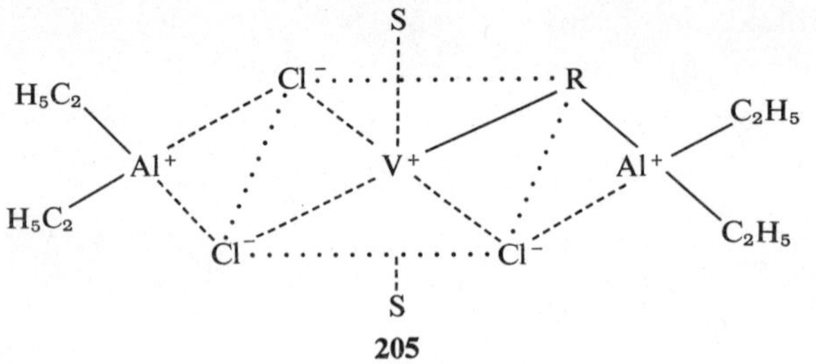

205

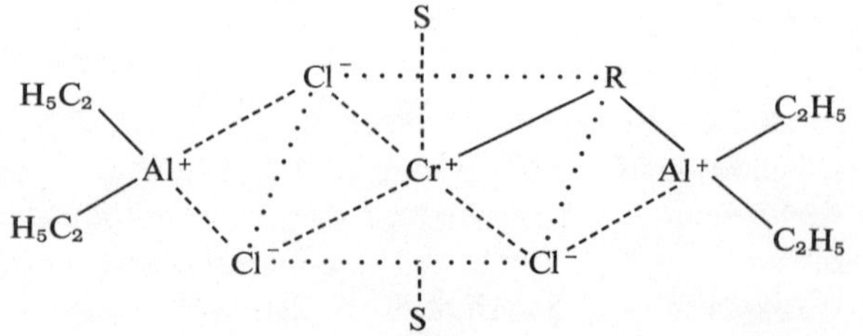

206

(S = solvent, R = Cl^- or C_2H_5)

The polymers formed have a broad molecular weight distribution, which is normally obtained with heterogeneous systems. It has been proposed [239] that the active catalytic sites in the homogeneous systems are the transition metal centers, chromium(II) or vanadium(II).

Natta *et al.* [240] have obtained alternate copolymers of ethylene and 2-butene by low-temperature (−30°C) polymerization of the monomers with catalyst systems consisting of VCl₄/Al(hexyl)₃ in *n*-heptane, and tris(2,4-pentanediono)vanadium(IV) and diethylaluminum chloride in toluene. The copolymer chains obtained have the repeating units indicated in **207**. An

$$-CH_2-CH_2-CH-CH-CH_2-CH_2-CH-CH-$$
$$\qquad\qquad\quad |\quad | \qquad\qquad\qquad\quad |\quad |$$
$$\qquad\qquad\quad CH_3\ CH_3 \qquad\qquad\qquad CH_3\ CH_3$$

207

infrared band at 13.2 mμ has been attributed to sequences of two adjacent methylenic (—CH₂—) groups [224]. Absorption bands between 13.6 and 13.9 mμ due to longer sequences of methylenic groups are absent. The copolymer is noncrystalline, as evidenced by X-ray diffraction studies, and it was found not to be possible to obtain copolymers with 2-butene contents of more than 50 mole %. Syndiotactic polymers of propylene with high steric regularity have been obtained by the use of tris(2,4-pentanediono)oxovanadium(IV) and diethylaluminum chloride in a hydrocarbon solvent. The activity of this catalyst can be maintained only at low temperatures (∼ −78°C) [217]. The EPR spectra of the catalyst solution obtained soon after mixing the components and after 3 hours, respectively, did not show signals indicating the presence of vanadium(III). At room temperature, however, a spectrum identical to that of vanadium(II) was observed, and the resulting catalyst was inactive for the polymerization of propylene. It was therefore concluded [241] that vanadium(II) is not the active species for the low-temperature polymerization of propylene. It has been suggested that an alkylvanadium(III) dichloride and aluminum alkyl complex is the active catalyst, as indicated by the color change of the vanadium tetrachloride solution on the addition of aluminum alkyl in *n*-heptane or toluene solutions [217].

The polymerization of cyclobutene [230a] in the presence of soluble tris-(2,4-pentanediono)–Cr(III) + Al(C₂H₅)₂Cl or tris(2,4-pentanediono)–V(III) + Al(C₂H₅)₂Cl at −30°C afforded the disyndiotactic polycyclobutene indicated by formula **208**. The use of heterogeneous VOCl₃ or tris(2,4-pentanediono)vanadium(IV) resulted in the formation of diisotactic polycyclobutene, **209**. The formation of the polycyclobutenomers seems to require the presence of a catalyst containing chloride ions coordinated either to aluminum or to a transition metal ion. Disyndiotactic and diisotactic polycyclobutenomers **208** and **209**, respectively, differ in their X-ray patterns and melting points. The melting point of **208** is about 150°C and that of **209**, about 210°C. The two polymers are, however, identical in their chemical composition and have similar molecular weights, the only difference being the configurations of the tertiary carbon atoms of the main chains.

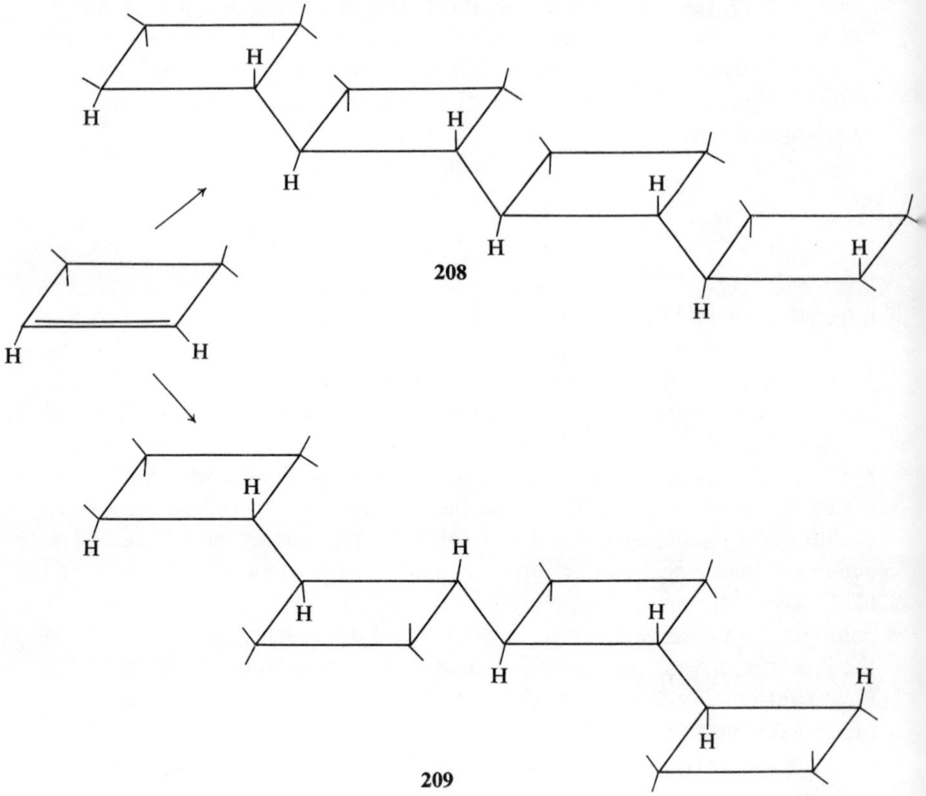

208

209

B. TRANSITION METAL ION CATALYSIS

Stereospecific catalysis of polymerization of cyclobutene can be carried out in water or alcohol solution with a ruthenium(III) chloride catalyst [224]. Catalysis by ruthenium(III) [224] results in the formation of both 1,4-*cis* and 1,4-*trans* isomers of polybutadiene (entries 17–20 in Table III) in relative amounts that are dependent on the solvent and reaction temperature. Cyclobutene as well as 3-methylcyclobutene are polymerized in aprotic solvents by ruthenium(III) chloride, probably by a ring cleavage reaction. The polymers obtained consist of either 1,4-butadiene or 1,4-pentadiene units with 1,2-butadiene units and isoprene units completely absent [224]. It has been proposed [224] that the cleavage of the 3-methylcyclobutene ring occurs at

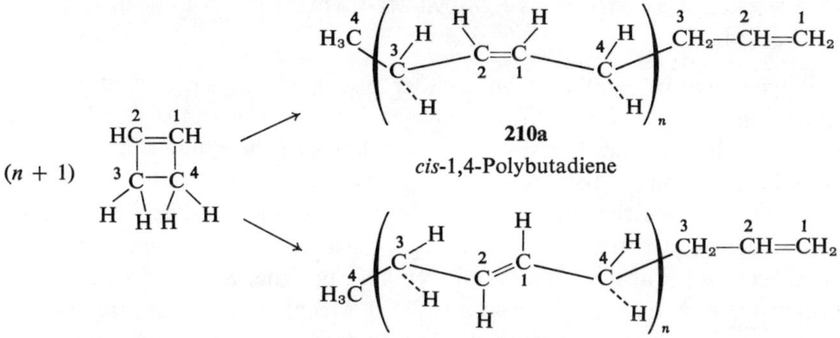

$(n + 1)$ $HC\overset{2}{=}\overset{1}{CH}$ $\overset{3}{C}-\overset{4}{C}$ H H H H

210a
cis-1,4-Polybutadiene

210b
trans-1,4-Polybutadiene

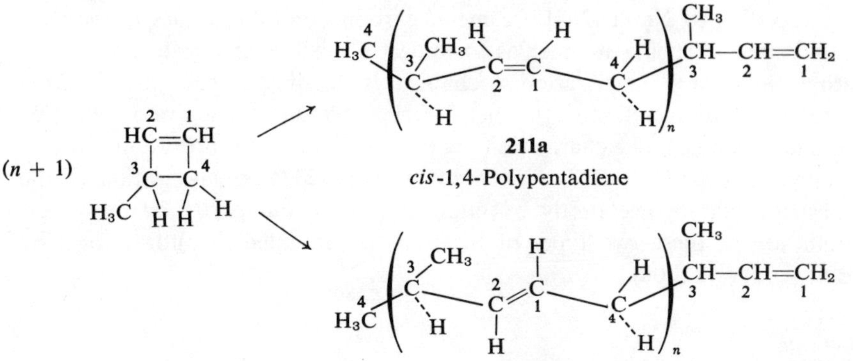

$(n + 1)$ $HC\overset{2}{=}\overset{1}{CH}$ $\overset{3}{C}-\overset{4}{C}$ H_3C H H H

211a
cis-1,4-Polypentadiene

211b
trans-1,4-Polypentadiene

carbon atoms 1 and 4 to give vinylidene end groups in the growing polymer chain. The relative tendencies of transition metal ions to catalyze ring cleavage of cyclobutenes increase in the order chromium(III) < vanadium(III) < molybdenum(VI) < titanium(III) and nickel(II) < rhodium(III) < ruthenium(III). An example of an exception to ring cleavage polymerization is the catalysis of the polymerization of cyclobutene by rhodium(III) chloride in aqueous sodium dodecylbenzene sulfonate emulsion at 50°C and at a monomer/rhodium(III) ratio of 200 to give diisotactic polycyclobutenomer [223]. The polymerization of cyclobutene by ring cleavage nevertheless takes place to a limited extent as a side reaction.

Polystyrene of high molecular weight has been obtained [222] in high yields by the rhenium pentachloride-catalyzed polymerization of styrene in benzene or ethylene dichloride solution. Silver perchlorate also catalyzes [221] polymerization of vinyl compounds in organic solvents. Kinetic and mechanistic aspects of these reactions are presented in the following section.

C. KINETICS AND MECHANISM OF METAL-CATALYZED POLYMERIZATION OF ALKENES

Polymerization of alkenes with either homogeneous Ziegler catalysts or transition metal complexes in homogeneous systems proceeds mostly by an ionic coordination mechanism [204]. Activation of the π bond through pre-equilibrium complex formation usually takes place before the subsequent propagation steps that involve multiple insertion of an alkene in a metal–ligand bond. In certain cases, especially those involving 2,4-pentanediono complexes in the absence of aluminum alkyls, polymerization of alkenes may also proceed by free-radical processes [242]. Mechanisms of such free-radical reactions are, however, not considered in this section because of lack of information about the rate-determining steps and the reactive intermediates involved.

According to Natta [204] if a metal–carbon bond dissociates in the activated state in such a manner as to impart a positive charge to the carbon atom, then the polymerization mechanism is classified as cationic. If on the other hand, dissociation of the metal–carbon bond in the activated complex results with a negative charge residing primarily on the carbon atom, then an anionic mechanism takes over. In some cases [232] polymerization of the substrate takes place partly by one mechanism and partly by the other. Initiation of these two types of reactions is presented in outline form by Eqs. (205) and (206).

Cationic

$$M^{\delta+}\!-\!X^{\delta-} + CH_2\!=\!CHR \longrightarrow M\!-\!CH_2\!-\!\overset{\overset{\displaystyle H}{\displaystyle |}}{\underset{\underset{\displaystyle R}{\displaystyle |}}{C}}{}^{\delta+}\!\cdots X^{\delta-} \qquad (205)$$

Anionic

$$M^{\delta+}\!-\!R^{\delta-} + CH_2\!=\!CHR \longrightarrow M^{\delta+}\!\cdots CH_2^{\delta-}\!-\!\overset{\overset{\displaystyle H}{\displaystyle |}}{\underset{\underset{\displaystyle R}{\displaystyle |}}{C}}\!-\!R \qquad (206)$$

R = alkyl, aryl

In reactions that take place by a cationic mechanism, polarity of the metal–carbon bond and the stability of the resulting carbonium ion play important parts in the polymerization process. In reactions of this type extensive isomerization also takes place by prototropic shifts in the direction of more stable carbonium ions. Because of this factor, cationic polymerization of vinyl monomers are nonstereospecific, unless conducted at low temperature ($-100°$ to $40°C$), under which conditions prototropic shifts may be very slow [204].

The mechanism proposed [222] for the rhenium pentachloride-catalyzed cationic polymerization of styrene in benzene solution is illustrated by reactions (207)–(210).

Initiation

$$\text{ReCl}_5 + \text{C}_6\text{H}_5\text{CH}{=}\text{CH}_2 \xrightleftharpoons{K_1} \begin{array}{c} \text{C}_6\text{H}_5\text{CH}{\overset{\shortmid}{=}}\text{CH}_2 \\ \text{ReCl}_5 \end{array} \qquad (207)$$

$$\begin{array}{c} \text{C}_6\text{H}_5\text{CH}{\overset{\shortmid}{=}}\text{CH}_2 \\ \text{ReCl}_5 \end{array} + \text{C}_6\text{H}_5\text{CH}{=}\text{CH}_2 \xrightleftharpoons{K_2} \left[\begin{array}{c} \text{C}_6\text{H}_5\text{CH}{\overset{\shortmid}{=}}\text{CH}_2 \\ \text{ReCl}_4 \\ \text{C}_6\text{H}_5\text{CH}{\overset{\shortmid}{=}}\text{CH}_2 \end{array} \right]^{+} \text{Cl}^{-} \qquad (208)$$

Propagation

$$\text{Sty}\text{---}\text{ReCl}_4\text{---}\{\text{C}_6\text{H}_5\text{---}\text{CH}(\text{CH}_3)\text{---}\text{CH}{=}\text{CH}(\text{C}_6\text{H}_5)\}^{+}\,\text{Cl}^{-} + \text{Sty}$$

$$\longrightarrow \text{Sty}\text{---}\text{ReCl}_4\text{---}\{\text{C}_6\text{H}_5\text{---}\text{CH}(\text{CH}_3)\text{---}\text{CH}_2\text{---}\text{CH}(\text{C}_6\text{H}_5)\text{---}\text{CH}{=}\text{CH}(\text{C}_6\text{H}_5)\}^{+}\,\text{Cl}^{-} \qquad (209a)$$

$$\longrightarrow \text{Sty}\text{---}\text{ReCl}_4\text{---}\{\text{C}_6\text{H}_5\text{---}\text{CH}(\text{CH}_3)\text{---}[\text{CH}_2\text{---}\text{CH}(\text{C}_6\text{H}_5)\text{---}]_{n-1}\text{---}\text{CH}{=}\text{CH}(\text{C}_6\text{H}_5)\}^{+}\,\text{Cl}^{-}$$
$$+ \text{Sty} \quad (209b)$$

$$\longrightarrow \text{Sty}\text{---}\text{ReCl}_4\text{---}\{\text{C}_6\text{H}_5\text{---}\text{CH}(\text{CH}_3)\text{---}[\text{CH}_2\text{---}\text{CH}(\text{C}_6\text{H}_5)\text{---}]_{n}\text{---}\text{CH}{=}\text{CH}(\text{C}_6\text{H}_5)\}^{+}\,\text{Cl}^{-} \quad (209c)$$

Termination

$$\text{Sty}\text{---}\text{ReCl}_4\text{---}\{\text{C}_6\text{H}_5\text{---}\text{CH}(\text{CH}_3)\text{---}[\text{CH}_2\text{---}\text{CH}(\text{C}_6\text{H}_5)\text{---}]_{n}\text{---}\text{CH}{=}\text{CH}(\text{C}_6\text{H}_5)\}^{+}\,\text{Cl}^{-}$$

$$\longrightarrow \text{Sty}\text{---}\text{ReCl}_5\text{---}\{\text{C}_6\text{H}_5\text{---}\text{CH}(\text{CH}_3)\text{---}[\text{CH}_2\text{---}\text{CH}(\text{C}_6\text{H}_5)\text{---}]_{n}\text{---}\text{CH}{=}\text{CH}(\text{C}_6\text{H}_5)\} \qquad (210)$$

Evidence for the formation of a π complex in the preequilibrium steps and in the subsequent propagation steps has been provided by the spectra of the various complex rhenium(V) species in solution.

Cationic polymerization reactions of styrene and of methyl methacrylate in benzene solution at 70°C are catalyzed by silver perchlorate. In the case of styrene the rate of polymerization is proportional to the first power of the silver(I) concentration at concentrations of silver(I) below 10^{-2} M [221]. Above this concentration, the rate of polymerization of styrene becomes proportional to the square of the silver(I) concentration. Second-order kinetics has been observed with respect to monomer for both styrene and methyl methacrylate polymerizations [227]. These kinetic results are explained on the basis of the assumption that a silver perchlorate monomer complex is formed in a preequilibrium step, followed by combination with a second monomer molecule in a cationic polymerization step. Complexes have been reported with both 2:1 and 1:1 molar ratios of silver(I) to styrene depending on the concentration of silver(I) in solution, as indicated by the corresponding equilibria, (211) and (212). The reaction mechanism proposed for the polymerization reaction at low silver(I) concentration is illustrated by reaction sequence (213).

The Ag(I)–styrene complex **212** formed in preequilibrium step (211) adds a second mole of styrene in **213** in a rate-determining step involving heterolytic fission of a C—H bond of the second styrene molecule and hydride transfer to the CH_2 group of the coordinated styrene molecule. The hydride transfer step seems to be assisted by metal ion polarization of the double bond in the complexed styrene. This mechanism seems to be in accord with a moderate deuterium isotope effect (~ 2).

$$Ag^+ + C_6H_5CH\text{=}CH_2 \underset{}{\overset{K_1}{\rightleftharpoons}} \begin{array}{c} C_6H_5CH\text{=}CH_2 \\ \vdots \\ Ag^+ \end{array} \tag{211}$$

212

$$Ag^+ + Ag(C_6H_5CH\text{=}CH_2)^+ \overset{K_2}{\rightleftharpoons} [Ag_2(C_6H_5CH\text{=}CH_2)]^{2+} \tag{212}$$

(Values of K_1 and K_2 at 70°C are 18 and 10, respectively.)

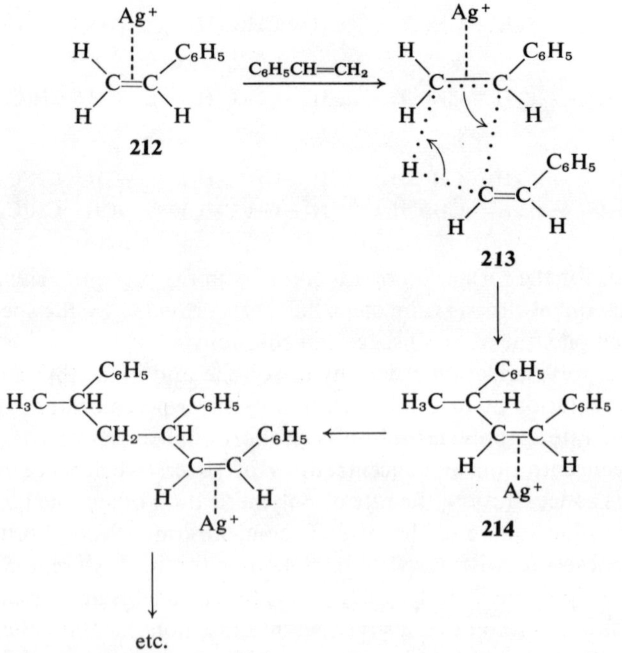

(213)

etc.

Another possible mechanism that cannot be ruled out is the participation of silver(I) in the hydride transfer step by the formation of intermediate hydride complexes. A possible course for such a mechanism is reaction sequence (214).

Similar mechanistic schemes may be used to explain the cationic polymerization of methyl methacrylate. The intermediates and products formed would be similar to those formed by styrene, with $-CH_3$ and $-COOCH_3$ replacing the $-H$ and $-C_6H_5$ attached to the α-carbon of the styrene residue.

Polymerization of alkenes with Ziegler catalysts generally proceeds by a metal ion-coordinated anionic mechanism. An anionic mechanism was initially proposed [204] to account for the stereospecific polymerization of α-olefins in the presence of heterogeneous catalysts. Similar mechanisms also apply to the homogeneous catalysis of Ziegler-type polymerization [225,243].

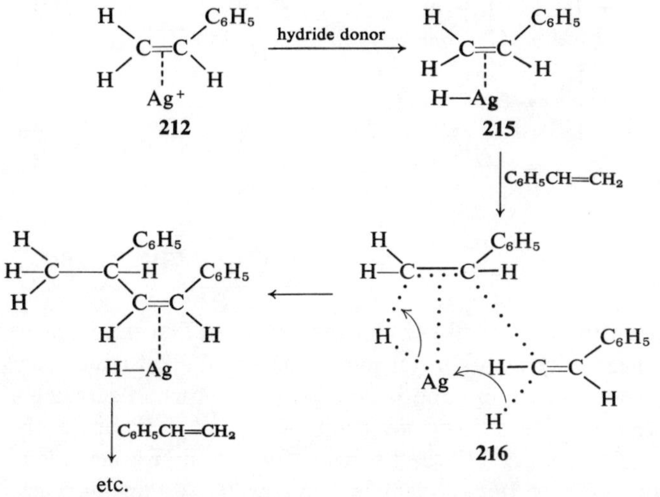

(214)

Of the two metals involved in Ziegler catalysis, a transition metal ion and an aluminum alkyl in a mixed-ligand complex, it was originally thought that the active center of the catalyst is aluminum ion. Later, Carrick [244] and Karol and Carrick [245] reported that a change in the transition metal center in the series hafnium(IV), zirconium(IV), titanium(IV), and vanadium(IV), in catalysts containing the same aluminum alkyl compound strongly influenced the rate of copolymerization of propylene and ethylene. The rate increases with decreasing electronegativity of the metal ions in the series vanadium(IV) > titanium(IV) > zirconium(IV) > hafnium(IV). On the other hand, change in the metal alkyl from triisobutylaluminum to diphenylzinc, di-n-butylzinc, or methyltitanium trichloride, in combination with a common transition metal ion, gave the same reactivity in all cases. This observation [244,245] strongly indicates that the transition metal ion is the active center in Ziegler catalysis. Matlack and Breslow [227] provided strong evidence to the effect that polymerization takes place on a transition metal ion center alone by studying the polymerization of ethylene and propylene in the presence of a catalyst system composed exclusively of transition metal compounds. The combinations used were free transition metal with alkyl halide, transition metal hydride, and transition metal with transition metal halide. For titanium-(IV), vanadium(IV), and zirconium(IV) the polymers isolated were characteristic of those obtained with Ziegler catalysts [241,246] and different from the polymeric products of the free-radical reaction. Accordingly it has been proposed that a metal alkyl is formed *in situ* from either a free metal and alkyl halide or a metal hydride and olefin [Eq. (215)] and that olefin insertion reactions occur by a coordinated anionic mechanism as indicated in Eq. (216).

$$Ti + C_2H_5Br \longrightarrow Ti^+—C_2H_5 + Br^- \qquad (215)$$
$$\underset{217}{}$$

$$TiC_2H_5^+ + C_2H_4 \longrightarrow \left[\begin{array}{l} CH_2 \\ \| \text{-----}Ti^{\delta+}—^{\delta-}C_2H_5 \\ CH_2 \end{array}\right]^+ \longrightarrow$$
$$\underset{218}{}$$

$$H_2C\cdots\cdots Ti^{2+}$$
$$H_2C\cdots\cdots C_2H_5 \qquad (216)$$

$$Ti—CH_2CH_2CH_2CH_3^+, Br^-$$
$$\underset{219}{}$$

A low-valence transition metal alkyl **217** is formed in the first step because of the low electronegativity of the metal ion in the metal alkyl under consideration, and the metal–carbon bond is polarized, with carbon carrying a partial negative charge. The anionic insertion of the alkyl group in the polarized metal–olefin complex **218** results in the formation of a new metal alkyl **219**. The process of insertion of olefin is then repeated, producing a new metal alkyl that propagates anionic olefin insertion in growing polymer chains. The presence of aluminum alkyls in Ziegler catalysts thus helps in the formation of a low-valent transition metal ion by reductive alkylation [225, 236,247]. The polymerization rate is independent of the aluminum alkyl concentration but is proportional to the concentration of the transition metal catalyst and the partial pressure of the olefin.

Cossee [243,246,248] has employed molecular orbital concepts in his treatment of mechanisms of Ziegler catalysis. A transition metal ion with empty or nearly empty d orbitals and with at least one alkyl group present in its coordination sphere may serve as an active catalytic center in homogeneous or heterogeneous Ziegler catalysis. A representative Ziegler catalyst with titanium(III) as the active center is illustrated in reaction sequence (217). The group X in complex **220** may be a simple halogen donor or a halogen bridging group, attached to the aluminum alkyl co-catalyst (not shown in complex **220** for clarity).

A variety of metal ions of the first, second, and third groups of the periodic table may serve as co-catalysts in Ziegler systems. Satisfactory co-catalyst metal ions have comparatively small ionic radii: Li^+ (0.68 Å), Be^{2+} (0.35 Å),

Al^{3+} (0.51 Å), Zn^{2+} (0.74 Å), and Mg^{2+} (0.66 Å), and as such may modify the electronegativity of the transition metal ion by coordination through bridging halogen groups. The co-catalyst also alkylates the transition metal ion and prevents its further reduction to the metallic state.

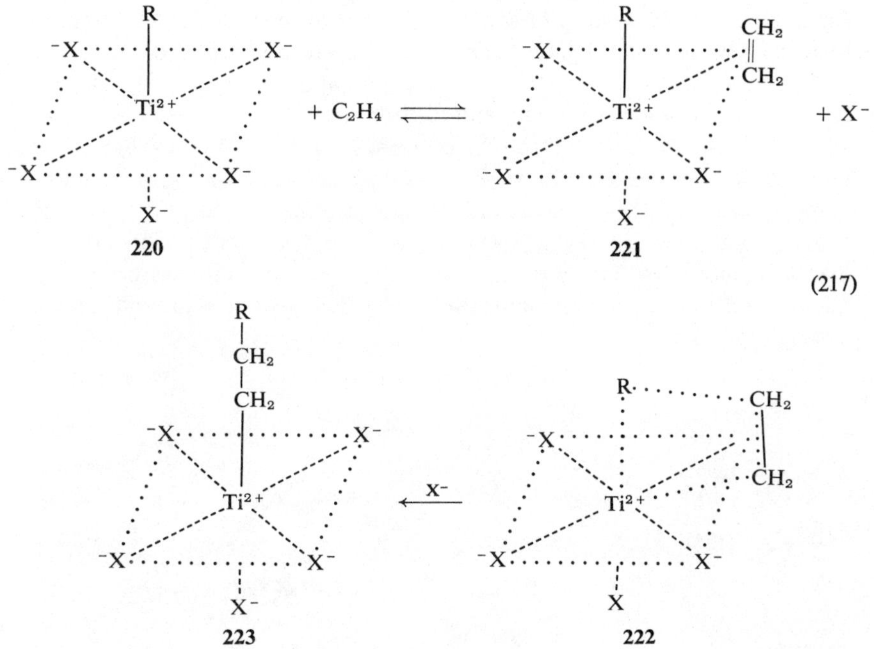

$$
\begin{array}{ccc}
\textbf{220} & + \; C_2H_4 \rightleftharpoons & \textbf{221} \quad + \; X^-
\end{array}
$$

(217)

$$
\textbf{223} \quad \xleftarrow{\;\; X^- \;\;} \quad \textbf{222}
$$

In the first step of any Ziegler type catalytic reaction, the transition metal ion complex (e.g., **220**) forms a π complex with an incoming monomer molecule. The formation of this π complex **221** with the olefin increases the distance between filled bonding and empty antibonding levels of a transition metal ion such as platinum(II) that has filled t_{2g} levels [249].* In the case of transition metal ions such as titanium(III) with empty, or nearly empty, t_{2g} levels, coordination of the metal ion with an olefin reduces the distance between the highest occupied bonding levels of the transition metal alkyl and its empty t_{2g} levels (see Fig. 1, Chapter 1). This causes lability of the otherwise stable metal–alkyl bond that breaks homolytically into alkyl radicals by expulsion of one of the bonding electrons into the antibonding levels of the coordinated olefin through the metal t_{2g} orbitals. The alkyl radical then forms a bond with the carbon atom of the olefin that already has an electron in its antibonding level. Transfer of the alkyl group to the olefinic carbon atom and the formation of a new Ti–carbon bond are probably concerted processes

* Discussed in Chapter 1 of this book.

with very little net nuclear displacement, requiring primarily electronic rearrangements. Formation of the alkene complex **221** is reversible but the migration of R is irreversible [202] and is favored by the energy of saturation of the double bond (~ 10 kcal/mole). On the basis of crystallographic data in an olefin π complex reported by Alderman *et al.* [250], Cossee [248] has estimated that in the transfer of the alkyl group to the olefinic carbon atom, the distance between the negative carbon of the alkyl group and the nearest carbon atom of the alkene must contract from about 3.3 to 1.54 Å. The relative positions of the alkyl group and olefin molecule in complex **222** are thus very favorable for the alkyl transfer process. Formation of a new metal alkyl complex **223** vacates a position on the transition metal ion to form a new π complex with another molecule of the alkene. In each insertion step, the growing monomer chain thus oscillates between two coordination positions about the metal ion. This process of multiple insertion continues until subsequent chain-termination reactions take place. Such reactions are indicated by Eqs. (218)–(220).

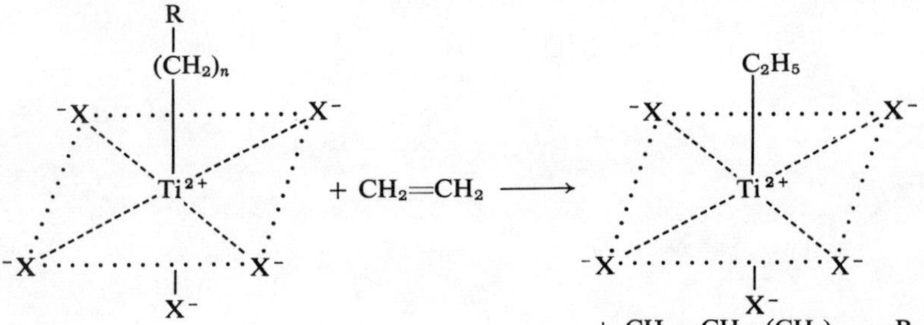

$$\begin{array}{c} R \\ | \\ (CH_2)_n \\ X\cdots|\cdots X \\ \diagdown Ti^{2+} \diagup \\ X\cdots|\cdots X \\ X^- \end{array} + CH_2{=}CH_2 \longrightarrow \begin{array}{c} C_2H_5 \\ | \\ X\cdots|\cdots X \\ \diagdown Ti^{2+} \diagup \\ X\cdots|\cdots X \\ X^- \end{array}$$

$$+ CH_2{=}CH{-}(CH_2)_{n-2}{-}R \qquad (218)$$

$$\begin{array}{c} R \\ | \\ (CH_2)_n \\ X\cdots|\cdots X \\ \diagdown Ti^{2+} \diagup \\ X\cdots|\cdots X \\ X^- \end{array} + RAlR_2 \longrightarrow \begin{array}{c} R \\ | \\ X\cdots|\cdots X \\ \diagdown Ti^{2+} \diagup \\ X\cdots|\cdots X \\ X^- \end{array}$$

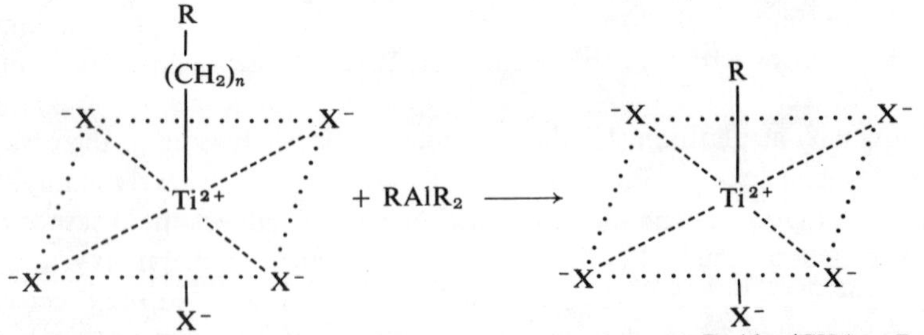

$$+ R_2Al{-}(CH_2)_n{-}R \qquad (219)$$

Termination of the growing polymer chain by exchange with a molecule of the monomer in reaction (218) is supported by the studies of Ikeda and Tsuchiya [251] on the polymerization of *cis*- and *trans*-CHD=CHD with TiCl$_4$ by Al(isobutyl)$_3$. Besides polymerization of deuterated olefin, *cis–trans* isomerization and hydrogen–deuterium exchange reactions were observed in this catalytic system. The deuterium exchange [251] reaction involves equilibrium conversion of the coordinated alkyl to the alkene indicated in Eq. (221).

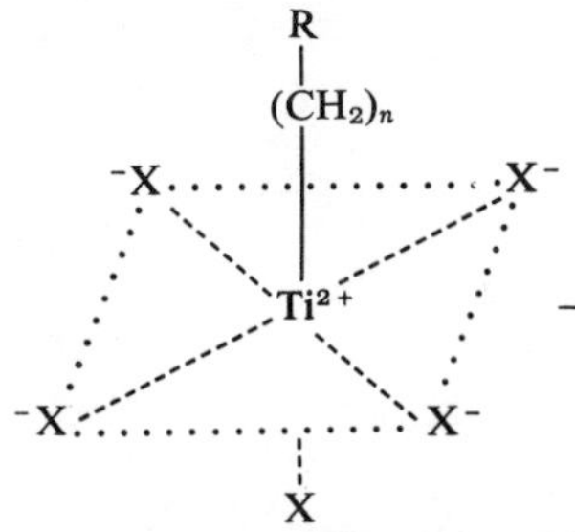

$$\text{Ti—H}^{2+} + 5\text{ X}^- + \text{CH}_2\!\!=\!\!\text{CH—(CH}_2)_{n-2}\text{—R} \tag{220}$$

$$\text{M—CH}_2\text{—CH}_3 + \text{CHD}\!\!=\!\!\text{CHD} \rightleftharpoons \text{M—CHD—CH}_2\text{D} + \text{CH}_2\!\!=\!\!\text{CH}_2 \tag{221}$$

In order that a transition metal ion be a good Ziegler catalyst, its d orbitals should be sufficiently diffuse to form a π bond with the incoming olefin. This condition restricts the catalytic activity to ions with low effective nuclear charge, a condition fulfilled by the catalytically active transition metal species that are reduced to a low-valent state in the Ziegler catalytic system [238]. Reduction of the charge on the metal ion reduces its electronegativity and helps to make the metal–alkyl σ bond sufficiently polar to encourage attack of the olefin by electron donation from the alkyl group. The metal alkyl formed by the interaction of aluminum alkyl and the transition metal ion should have considerable stability in the absence of the olefin, but should become kinetically labile in the presence of the coordinated alkene. These requirements are satisfied by transition metal ions containing from zero to three electrons in their d orbitals, and especially those of titanium(III) [238, 252] and vanadium(III) [237]. Metal ions of the iron and platinum groups, [Iron(II), cobalt(II), nickel(II), ruthenium(III) and rhodium(III)] that contain nearly filled d orbitals in their active oxidation states probably catalyze polymerization of olefins by a mechanism somewhat different from that described above.

III. Polymerization of Dienes

Stereospecific polymerization of butadiene can give rise to four different polymers [204], 1,2-isotactic, 1,2-syndiotactic, cis-1,4, and trans-1,4 as indicated by formulas 224–227.

Substituted polybutadienes may exist in isomeric forms analogous to those of polybutadiene itself. For example, isoprene (2-methylbutadiene) can form isomeric polymers with basic 1,2-, 3,4-, 1,4-cis, and 1,4-trans units, as is shown in structures 228–231.

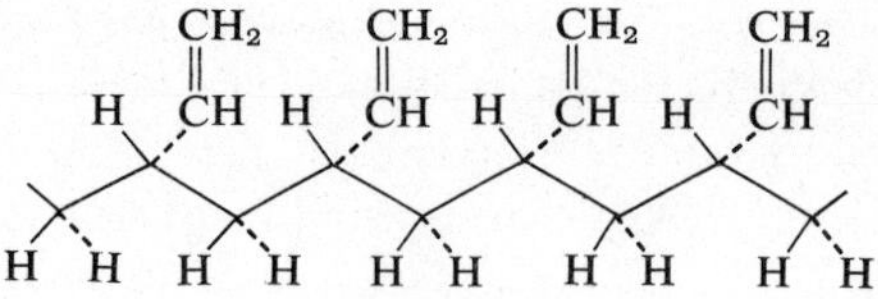

224

Isotactic 1,2-polybutadiene

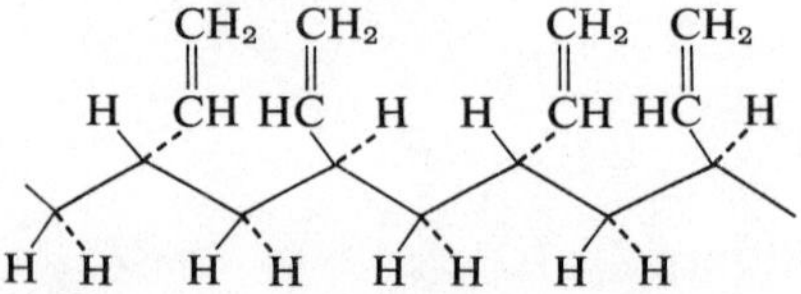

225

Syndiotactic 1,2-polybutadiene

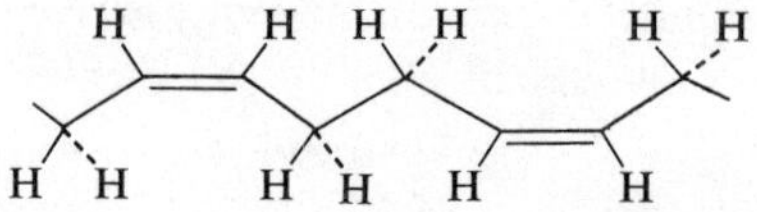

226

cis-1,4-Polybutadiene

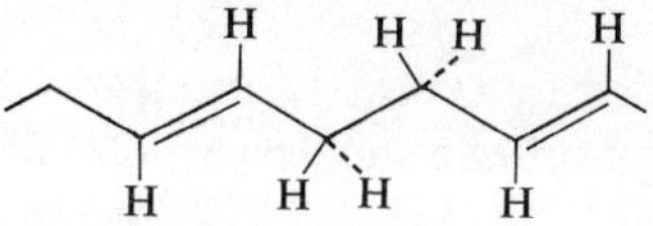

227

trans-1,4-Polybutadiene

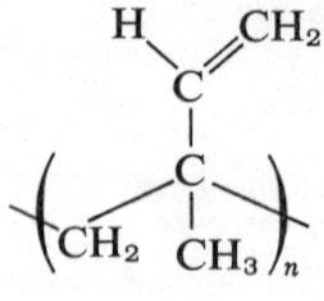

228

1,2-Unit

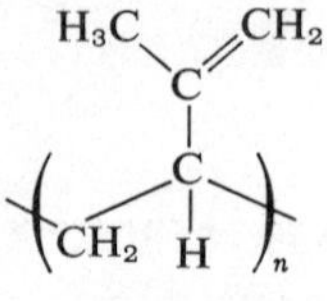

229

3,4-Unit

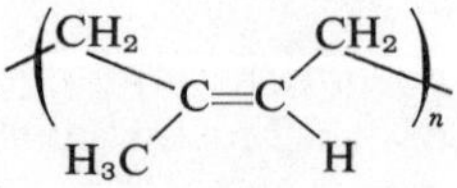

230

cis-1,4-Unit

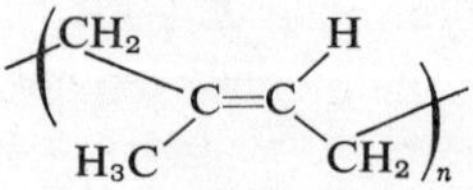

231

trans-1,4-Unit

Polybutadienes occurring in nature invariably have *cis*-1,4 and *trans*-1,4 structures, while 1,2-polybutadienes, which are actually substituted polyethylenes and thus are similar to polypropylene, have not been found to occur in nature [202]. Natural rubber molecules are composed of *cis*-1,2 polyisoprene units [216], in contrast to the naturally occurring polybutadienes. On the other hand, synthetic stereospecific polymers obtained by the catalytic polymerization of butadiene or its derivatives with either Ziegler catalysts or with other transition metal ion catalyst systems frequently have high *cis*-1,4-polybutadiene type content. Examples of catalysts used in polymerization of butadienes, and the types of polymers obtained, are listed in Table IV [204,231,236,242,253–269].

A. POLYMERIZATION BY ZIEGLER TYPE CATALYSTS

trans-1,4-Polybutadiene and *trans*-1,4-polyisoprene of high steric purity (99%) have been obtained by heterogeneous Ziegler catalysis [204] (see entries 1, 2, Table IV). The formation of the *trans* structure in these cases may be due to a staggered arrangement of the diene molecule adsorbed on the catalyst surface. Lehr and Moyer [253,254] obtained predominantly *cis*-1,4-polybutadiene with a heterogeneous *n*-butyllithium–titanium tetrahalide catalyst, at a molar ratio of Li/Ti of less than 2.5 (entries 3 and 4, Table IV). The structures of the polybutadienes formed were found to vary with the ratio of lithium to titanium. In the range of the catalyst ratio of 3/1 to 4/1, the polymers formed consisted predominantly of 1,2-polybutadienes, while for ratios of Li/Ti greater than 5/1, the polymers consisted of nonstereospecific mixtures of isomers characteristic of the *n*-butyllithium catalyst alone.

The structures obtained by the polymerization of butadiene [254] with a catalyst system consisting of aluminum alkyls and titanium(IV)halides (chloride, bromide, iodide) were found to depend on the nature of the halide ion. With a titanium(IV) chloride catalyst, the structure of the polymer obtained was found to be independent of the proportion of halide. With the iodide, however, the percentage of *cis*-1,4-butadiene polymer obtained decreases with increasing proportion of titanium(IV) iodide in the catalyst. These TiX_4–AlR_3 catalysts have also been used [255] for the polymerization of *d*-limonene to a low molecular weight polymer containing bicyclic saturated units of camphane or pinane (see entry 5, Table IV). The limonene moieties were found to be completely racemized during the polymerization process. Homogeneous catalysts containing titanium alkoxide and aluminum alkyl in benzene or *n*-heptane solution have been studied by Natta *et al.* [256,257,270]. The titanium alkoxides used include $Ti(n\text{-butoxide})_4$ and $Ti(isopropoxide)_4$. On mixing an aluminum alkyl with the titanium alkoxide, dark solutions are

TABLE IV

POLYMERIZATION OF DIENES BY MULTIPLE INSERTION REACTIONS

No.	Catalyst	Type of solvent	Substrate	Temp. (°C)	Products	Ref.
1	VCl_3 + AlR_3	Heterogeneous	1,3-Butadiene	15	trans-1,4-Polybutadiene	204
2	$VOCl_3$ + AlR_3	Heterogeneous	Isoprene	15	trans-1,4-Polyisoprene	204
3	TiX_4 + n-butyllithium					
	Li/Ti = 2.5	Heterogeneous	1,3-Butadiene	—	cis-1,4-Polybutadiene	253
	Li/Ti = 3–4	Heterogeneous	1,3-Butadiene	—	Amorphous 1,2-polymer	253
4	TiX_4 + AlR_3	Heterogeneous	1,3-Butadiene	—	cis-1,4-Polybutadiene	254
5	TiX_4 + AlR_3	Heterogeneous	d-Limonene	—	Low mol. wt. polymer	255
6	$Ti(OEt)_4$ + $Al(Et)_3$	Benzene	1,3-Pentadiene	15	cis-1,4-Isotactic polymer	256
7	$Ti(OR)_4$ + AlR_3	Toluene	Isoprene	15	Polymer with amorphous 3,4-units	257
8	$Ti(OR)_4$ + AlR_3	Toluene	Isoprene + 1,3-pentadiene	15	Copolymer with mixed 3,4-isoprene and cis-1,4-isotactic pentadiene units	257
9	$Ti(OR)_4$ + AlR_3	Benzene, n-heptane	1,3-Butadiene	15	1,2-Syndiotactic polybutadiene	257
10	$Ti(OR)_4$ + AlR_3	Benzene, n-hexane	1,3-Butadiene	15	Polybutadiene 1,2-syndiotactic, 10% + 1,2-amorphous, 90%	204
11	$MoO_2(OR)_2$ + AlR_3	Benzene, n-hexane	1,3-Butadiene	15	Polybutadiene 1,2-syndiotactic, 75% 1,2-amorphous, 25%	204
12	Bis(2,4-pentanediono)-MoO_2 + AlR_3	Benzene, n-hexane	1,3-Butadiene	15	Polybutadiene 1,2-syndiotactic, 75% 1,2-amorphous, 25%	204

No.	Catalyst	Solvent	Monomer	Temp.	Polymer	Ref.
13	Tris(2,4-pentanediono)-V(III) + AlR$_3$	Benzene, n-hexane	1,3-Butadiene	15	Polybutadiene 1,2-syndiotactic, 10% 1,2-amorphous, 90%	204
14	Tris(dipyridyl)-Cr(III) + AlR$_3$ (high Al/Cr ratio aged catalyst)	Benzene, n-hexane	1,3-Butadiene	15	Polybutadiene 1,2-isotactic 97–99%	204
15	Hexabenzonitrilo-Cr(III) + AlR$_3$ (high Al/Cr ratio aged catalyst)	Benzene, n-hexane	1,3-Butadiene	15	Polybutadiene 1,2-isotactic 97–99%	204
16	Cr(CO)$_n$(pyridine)$_m$ + AlR$_3$ (aged catalyst, $m + n = 6$)	Benzene, n-hexane	1,3-Butadiene	15	Polybutadiene 1,2-isotactic 97–99%	204
17	Tris(dimethylglyoximato)-Fe(II) + AlR$_3$	Benzene	1,3-Butadiene		Polybutadiene, cis-1,4	258
17a	Bis(dimethylglyoximato)-Fe(II) + AlR$_3$	Benzene	Isoprene		Polyisoprene 3,4-amorphous	258
18	Bis(dimethylglyoximato)-Ni(II) + AlR$_3$	Benzene	1,3-Butadiene		Polybutadiene, cis-1,4	258
18a	Bis(dimethylglyoximato)-Ni(II) + AlR$_3$	Benzene	Isoprene		Polyisoprene 3,4-amorphous	258
19	Tris(dimethylglyoximato)-Co(III) + AlR$_3$	Benzene	1,3-Butadiene		Polybutadiene, cis-1,4	258
19a	Tris(dimethylglyoximato)-Co(III) + AlR$_3$	Benzene	Isoprene		Polyisoprene 3,4-amorphous	258
20	Tris(2,4-pentanediono)-Co(III) + Al(C$_2$H$_5$)$_2$Cl	Benzene	1,3-Butadiene	25	Polybutadiene cis-1,4, trans-1,4, and 1,2-amorphous	259
21	Tris(2,4-pentanediono)-M(III) + Al(C$_2$H$_5$)$_2$Cl (M = Ti, V, Cr, Fe, Co)	Benzene	1,3-Butadiene	25	Polybutadiene cis-1,4, trans-1,4 and 1,2-amorphous	260

TABLE IV (*continued*)

No.	Catalyst	Type of medium	Substrate	Temp. (°C)	Products	Ref.
22	Bis(2,4-pentanediono)-M(II) + Al(C₂H₅)₂Cl (M = Mn, Fe, Ni)	Benzene	1,3-Butadiene	25	Polybutadiene cis-1,4, trans-1,4, and 1,2-amorphous	259
23	Bis(salicylaldehyde)-Co(II) + AlR₂Cl	Toluene	1,3-Butadiene	25	Polybutadiene cis-1,4, 99.3%	242
24	Bis(α-oxyacetophenone)-Co(II) + AlR₂Cl	Toluene	1,3-Butadiene	25	Polybutadiene cis-1,4, 98.7%	242 261
25	Bis(quinazarine)-Co(II) + AlR₂Cl	Toluene	1,3-Butadiene	25	Polybutadiene cis-1,4, 99.4%	242 261
26	Bis(orthovaline)-Co(II) + AlR₂Cl	Toluene	1,3-Butadiene	25	Polybutadiene cis-1,4, 99.4%	242 261
27	Bis(5-oxy-1,4-naphthaquinone)-Co(II) + AlR₂Cl	Toluene	1,3-Butadiene	25	Polybutadiene cis-1,4, 99.5%	242
28	Bis(8-oxyquinoline)-Co(II) + AlR₂Cl	Toluene	1,3-Butadiene	25	Polybutadiene cis-1,4, 99.3%	242 261
29	Bis(salicylaldehydeimine)-Co(II) + AlR₂Cl	Toluene	1,3-Butadiene	25	Polybutadiene cis-1,4, 99.3%	242 261
30	Bis(salicylaldehydeethylenediimine)-Co(II) + AlR₂Cl	Toluene	1,3-Butadiene	25	Polybutadiene cis-1,4, 99.0%	242 261
31	Bis(α-nitroso-β-naphthol)-Co(II) + AlR₂Cl	Toluene	1,3-Butadiene	25	Polybutadiene cis-1,4, 98.9%	242 261
32	Bis(mercaptobenzothiazole)-Co(II) + AlR₂Cl	Toluene	1,3-Butadiene	25	Polybutadiene cis-1,4, 99.4%	242 261
33	Bis(mercaptobenzoxazole)-Co(II) + AlR₂Cl	Toluene	1,3-Butadiene	25	Polybutadiene cis-1,4, 99.3%	242 261

No.	Catalyst	Solvent	Monomer	Temp. (°C)	Product	Ref.
34	Bis(mercaptobenzimidazole)-Co(II) + AlR₂Cl	Toluene	1,3-Butadiene	25	Polybutadiene cis-1,4, 99.5%	242
35	CoCl₂ + AlX₃	Benzene	1,3-Butadiene	25	Polybutadiene cis-1,4	261
36	Co(alcoholate) + Al(isobutyl)₂Cl	Benzene	1,3-Butadiene	25	Polybutadiene cis-1,4	262
37	Rh(III) chloride	Water	1,3-Butadiene	55	Polybutadiene trans-1,4, over 99%	263
38	Rh(III) chloride	Aqueous ethanol	1,3-Butadiene	135	Polybutadiene trans-1,4 over 99%	264
39	Rh(III) chloride + 1,3-cyclohexadiene	Aqueous ethanol	1,3-Butadiene	55	Polybutadiene, trans-1,4	265
40	Rh(III)	Water	1,3-Butadiene	50	Polybutadiene trans-1,4 99.5%	236
41	Rh(I)–olefin complex	Water	1,3-Butadiene	50	Polybutadiene trans-1,4 99.5%	236
42	Ir(III)	Water	1,3-Butadiene	50	Polybutadiene trans-1,4 99.5%	236
43	Ru(III) + (C₆H₅)₃P (1:6)	Water	1,3-Butadiene	50	Polybutadiene cis-1,4 68% trans-1,4 17% 1,2-15%	236
44	Ru(III) + (n-C₄H₉)₃P (1:10)	Water	1,3-Butadiene	50	Polybutadiene cis-1,4 15% trans-1,4 15% 1,2-70%	236
45	Pd(II)	Water	1,3-Butadiene	50	Polybutadiene trans-1,4 17% 1,2 83%	236
46	(NH₄)₂PdCl₄	Water	1,3-Butadiene	50	Polybutadiene trans-1,4 2% 1,2 98%	236
47	CoSiF₆	Water	1,3-Butadiene	50	Polybutadiene cis-1,4 88% trans-1,4 8% 1,2-4%	236
48	(C₂H₅)₄Ti	Benzene	1,3-Butadiene	50	Polybutadiene trans-1,2/1,4 5.1%	267
49	(CH₂=CH)₄Ti	Benzene	1,3-Butadiene	50	Polybutadiene trans-1,2/1,4 6.6%	268
50	(C₆H₅)₄Ti	Benzene	1,3-Butadiene	50	Polybutadiene trans-1,2/1,4 5.4%	268
51	(CH₂=CH)₃Cr	Benzene	1,3-Butadiene	50	Polybutadiene trans-1,2/1,4 1.6%	268
52	(CH₂=CH)₂Fe	Benzene	1,3-Butadiene	50	Polybutadiene trans-1,2/1,4 4.2%	268
53	(CH₂=CH)₂Co	Benzene	1,3-Butadiene	50	Polybutadiene trans-1,2/1,4 3.7%	268
54	(CH₂=CH)₂Ni	Cyclohexane	1,3-Butadiene	—	Polybutadiene trans-1,2/1,4 0.40%	269
55	π-(C₃H₅)₃Co	Cyclohexane	1,3-Butadiene	—	Polybutadiene trans-1,4	269
56	π-(C₃H₅)₂CoI	Cyclohexane	1,3-Butadiene	—	Polybutadiene trans-1,4	269
57	π-(C₃H₅)₃Cr	Cyclohexane	1,3-Butadiene	—	Polybutadiene 1,2	269

obtained and ethane is evolved [256]. The solutions, however, remain homogeneous at all ratios of Al/Ti. The evolution of ethane from the catalyst solution was attributed to formation of a titanium alkyl that is then cleaved with the formation of a low-valent titanium species and an ethyl radical. The valence state of titanium in the mixed aluminum–titanium complex is not known but there is indication that in the complex there is at least one alkoxy group still coordinated to titanium. This catalyst system polymerizes butadiene [204,257] to amorphous and crystalline (1,2-syndiotactic) polymers (entries 9–10, Table IV). Isoprene gave a completely amorphous polymer (Table IV, No. 7) with 3,4-units under the same conditions [257]. Both *cis*- and *trans*-1,3-pentadiene afforded polymers with predominant *cis*-1,4-isotactic structure (see entry 6, Table IV), as confirmed by X-ray and infrared spectroscopic studies [256].

Optically active *cis*-1,4-poly(1,3-pentadiene) has been obtained from triethylaluminum and titanium tetra-*l*-menthoxide. A mixture of (+)tris(2-methylbutyl)aluminum and titanium tetrabutoxide yielded *cis*-1,4-poly(1,3-pentadiene) having no optical activity. The fact that only optically active groups bound to titanium confer optical activity to the *cis*-1,3-polymer of 1,3-pentadiene lends support to the conclusion stated above that the active site for the polymerization of the diene is the transition metal ion. Copolymerization of isoprene and 1,3-pentadiene (No. 8, Table IV) yielded a polymer with mixed isoprene 3,4-units and *cis*-1,4-pentadiene moieties [257]. The difference in the polymers obtained from butadiene and 1,3-pentadiene with the use of a common $AlR_3 + Ti(OR)_4$ catalyst has been attributed to the difference in the nature of the π complex formed between monomer and the transition metal center [257]. It has been suggested that butadiene and isoprene are coordinated only through one double bond, whereas 1,3-pentadiene coordinates to the metal ion through both double bonds. The reason for this difference in coordination of the double bonds in butadiene and 1,3-pentadiene on titanium center is not clear. It is seen that 1,2-polymers of butadiene may also be obtained with catalysts such as molybdyl alkoxides with AlR_3 and tris(2,4-pentanediono)vanadium(IV) with AlR_3 (entries 11–13, Table IV). Maximum yields of the crystalline 1,2-syndiotactic polymer of butadiene have been obtained from the molybdyl catalyst [204]. The use of aged chromium(II) catalysts (entries 14–16, Table IV) results in the formation of a pure (99%) 1,3-isotactic polybutadiene.

B. MODIFIED ZIEGLER CATALYSIS

Soluble Ziegler catalysts containing dimethylglyoxime complexes of iron(II), nickel(II), and cobalt(II), combined with aluminum alkyls, have been used [258] to polymerize isoprene and butadiene (Table IV, entries 17–19a).

For the polymerization of butadiene, bis(dimethylglyoximato)iron(II) has been found to be a better catalyst than the corresponding cobalt(II) and nickel(II) complexes. The catalytic activities of dimethylglyoximino complexes were found to be higher than the complexes obtained from 2,4-pentanedione. The tris(2,4-pentanediono) complex of cobalt(III) on reaction with diethylaluminum chloride [259] is reduced to the corresponding cobalt(II) complex, as indicated by the fact that there is an increase in the absorption peak at 600–700 mμ, attributed to cobalt(II) ion, when the concentration of diethylaluminum chloride is increased. It has been proposed that the catalyst formed by the reaction of tris(2,4-pentanediono)cobalt(III) with diethylaluminum chloride is $CoCl_2(ClAlR_2)_2$, a chloride-bridged (μ-tetrachloro) polynuclear complex. This catalytic cobalt(II) system has been used by Natta et al. [271] in the stereospecific polymerization of butadiene to a cis-1,4-polymer.

In the polymerization of butadiene by the catalyst system consisting of tris(2,4-pentanediono)-Co(III)-Al(C_2H_5)$_3$ and AlX$_3$ (X = Cl$^-$, Br$^-$, I$^-$) (see Table IV, No. 20), the yield and the microstructure of the polybutadiene is greatly affected by the halide ion employed and by the molar ratio Al-(C_2H_5)$_3$/AlX$_3$ [259]. When the molar ratio is 1:1 and X = Cl$^-$, polymer yields greater than 90% with predominantly cis-1,4-structure were obtained. The use of AlBr$_3$ with the same ratio of AlEt$_3$/AlX$_3$ decreases the yield of the polymer to 80%, and the polymer distribution changes to 77% cis-1,4; 17% trans-1,4; and 6% 1,2-. When AlI$_3$ was used, the yield of polybutadiene was greater than 90% overall, with 50% cis-1,4; 19% trans-1,4; and 31% 1,2-. The catalytically active species in such solutions are most probably either $Co(ClAlR_2)_n$ or $Cl_nCo(ClAlR_2)_m$.

The effect of various 2,4-pentanediono complexes of titanium(III), cobalt-(III), iron(III), chromium(III), vanadium(III) manganese(II), and nickel(II) on the stereoregularity of the polymer obtained from butadiene at room temperature has been reported by Matsuzaki and Yasukawa [259,260] (Table IV, Nos. 21,22). The highest yield of cis-1,4-polymer was obtained with 1:1 and 2:1 ratios of Al(C_2H_5)$_3$/AlX$_3$. Vanadium(III) and chromium(III) catalysts produce polymers having comparatively higher 1,2-contents in good yields. The yields of cis-1,4-polymer decrease in the sequence nickel(II) > cobalt(III) > titanium(III) > iron(III) > manganese(II) > chromium(III) > vanadium(III). Yields of trans-1,4-polymer increase in the order nickel(II) < cobalt(III) < titanium(III) < iron(III) < manganese(III) < chromium-(III) < vanadium(III). The overall yield of the polymer increases in the order nickel(II) < cobalt(III) < titanium(III) < vanadium(III) < chromium(III) < iron(III) < manganese(II). In all these cases, the transition metal ion catalyst was invariably reduced to a low valence state during the course of polymerization. The yield and stereoregularity of the polymer may also be

126   6. MULTIPLE INSERTION REACTIONS

changed by other factors, such as change in the halide ion, the ratio of transition metal ion to aluminum compound, and addition of water or basic reagents to the reaction system. Bawn and Ledwith [235] have discussed the effect of transition metal ions on the polymerization of butadiene in considerable detail.

Takahashi and Kambara [261] and Kambara *et al.* [242] have polymerized isoprene and butadiene with Ziegler catalysts composed of alkylaluminum halides and chelate compounds of cobalt(II) having ligands containing (O,O), (O,N), (N,N), (O,S), (N,S), and (S,S) donor atoms (entries 23–34, Table IV). In most cases, the stereopurity of the *cis*-1,4-polymer was more than 99%. The *trans*-1,4-polymer was the only other isomer and was present in less than 1% concentration in the system. The reaction proceeds in aromatic or halohydrocarbon solvents even at very low concentrations of the cobalt(II) chelate catalyst, the reaction mixture remaining homogeneous throughout the polymerization process.

An interesting homogeneous catalyst for the polymerization of butadiene has been obtained by the combination of cobaltous chloride with aluminum chloride or bromide in benzene or cyclohexane solvents [262] (No. 35, Table IV). Cobaltous chloride is very insoluble in hydrocarbon solvents, but in the presence of aluminum halides, a soluble green complex is formed. The presence of charge-transfer bands at 300–360 mμ in benzene solution (but not in cyclohexane) indicates possible coordination of one or more benzene rings to cobalt(II). For the cobaltous complexes in benzene and cyclohexane solutions, the structures **232** and **233** have been suggested [262].

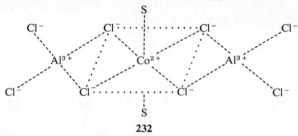

232

Probable structure of Co(AlCl$_4$)$_2$ in benzene solution [262]

cis-1,4-Polybutadiene has been obtained [263] from a catalytic system consisting of (CoCl$_2$, *n*-C$_2$H$_5$OH) and diisobutylaluminum chloride with molar ratios of cobalt/aluminum of about 200 (No. 36, Table IV). Water is also required, the maximum polymerization rate being observed at a molar ratio of water to aluminum of 1:10. In the absence of water, the insoluble copper–aluminum catalyst does not bring about the formation of high polymer, an oligomer of butadiene being formed. The important role of water in this catalyst system is not yet fully understood.

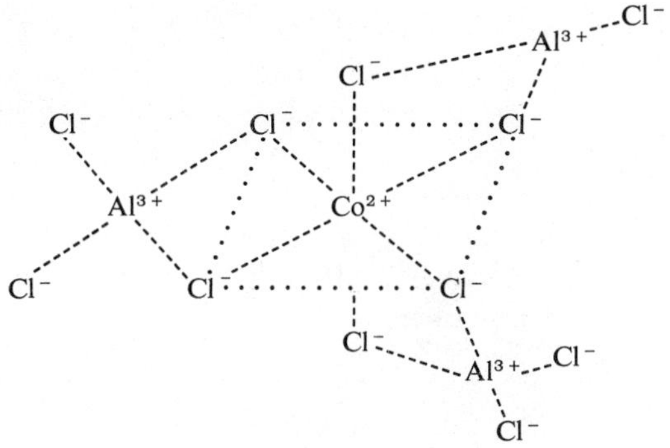

233

Probable structure of $Co(Al_3Cl_{11})$ in cyclohexane solution [262]

C. Non-Ziegler Type Transition Metal Catalysts

Transition metal ions (i.e., transition metal salts), especially those belonging to the platinum metal group, produce stereospecific polymerization of butadiene in homogeneous systems without added aluminum alkyl. Rhodium-(III) catalyzes stereospecific polymerization of butadiene to an exclusive *trans*-1,4-polymer (Nos. 37,38, Table IV) in aqueous or ethanol solution even in the presence of air [264,266]. The salts $Rh(NO_3)_3$ and $RhCl_3$ have been found to be effective polymerization catalysts in ethanol and aqueous solutions, respectively [266]. Intrinsic viscosities as high as 1.0 have been obtained for the butadiene polymer prepared with a $RhCl_3$ catalyst in an aqueous emulsion containing sodium laurylbenzene sulfonate [265]. Increasing the concentration of rhodium(III) beyond a certain range caused a lowering of the molecular weight of the polymer. In all cases, *trans*-1,4-polybutadiene separated from solution because of its highly crystalline nature and insolubility in cold benzene solution.

Teyessie and Dauby [267] have studied the polymerization of butadiene at 55°C by the addition of 1,3-cyclohexadiene to a rhodium(III) catalyst (Table IV, No. 39). The accelerating effect of the polyene reducing agents in the rhodium(III)-catalyzed butadiene polymerization was attributed to the participation of a lower-valent rhodium species, possibly a rhodium(I) or a rhodium(I) hydride complex [267]. Bawn and co-workers [272] proposed that the active species in the long-chain polymerization of butadiene is a π-alkylrhodium(III) hydride rather than Rh(I) π-allylic complexes. The stereospecific polymerization of butadiene to the *trans*-1,4-polymer in aqueous solution may be visualized as taking place by reaction sequence (222) [273].

Initial combination of rhodium(III) with butadiene and water results in the addition of a hydroxide ion to the ligand and formation of a negative π-allyl group which coordinates to the metal ion to give complex **234**. The negative π-allyl group then attacks an adjacent coordinated butadiene molecule (see **235**) and a new π-allyl group is formed. In complex **236** simultaneous coordination of the new terminal π-allyl group and the isolated double bond of the first coordinated butadiene is possible and may be one of the driving forces for the reaction. The steric requirement of chelate formation imposes stereospecificity in the polymer chain and results in the formation of an exclusively *trans* 1,4-polybutadiene.

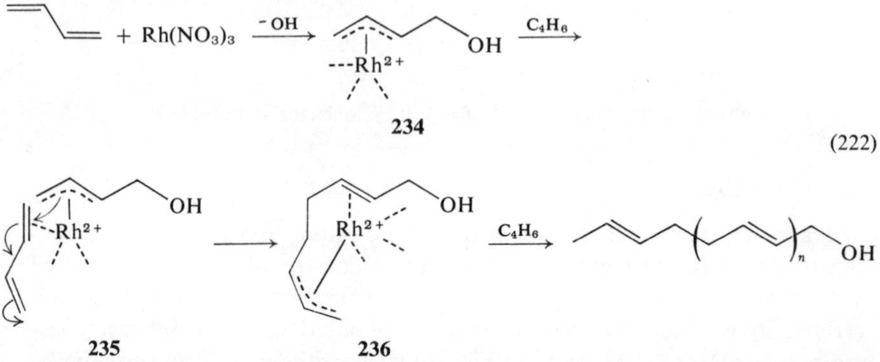

234

(222)

235   **236**

Polymerization of butadiene by organotitanium catalysts (Nos. 48–50, Table IV) in hydrocarbon or tetrahydrofuran solvents gave polybutadienes rich either in *trans*-1,4-moieties or 1,2-units, depending on the organo group bound to titanium [268]. Use of organo derivatives of chromium, iron, cobalt, and nickel resulted in the formation of low molecular weight polybutadienes accompanied by the cyclic oligomer, *trans,trans,trans*-1,5,9-cyclododecatriene (Table IV, entries 51–54).

Use of π-allyl derivatives of cobalt or nickel as catalysts in hydrocarbon solvents resulted in the stereospecific polymerization of butadiene [269]. Tris(π-allyl)cobalt gave *trans*-1,4-polybutadiene, whereas bis(π-allyl)cobalt iodide yielded the *cis*-1,4-isomer. Tris(π-allyl)chromium catalysis produced the 1,2-isomer (entries 55–57, Table IV).

It is of interest to compare the catalytic properties of various transition metal ions of the platinum group that have been found [236] active in the polymerization of butadiene (Table IV, Nos. 40–46). While rhodium(I) and iridium(I) form exclusively *trans*-1,4-polymers, ruthenium(III) gives a mixture of all possible polymers, *trans*-1,4; *cis*-1,4; and 1,2. On the other hand, catalysis by palladium(II) yields predominantly 1,2-polymers. It may be seen from Table IV that with palladium(II) the yield of 1,2-polymer depends on the coordinated chloride ligands groups; $PdCl_2$ gave 83% 1,2- and 17% *trans*-1,4-isomer, whereas $PdCl_4^{2-}$ yielded 98% 1,2- and only 2% *trans*-1,4-polymer.

Factors that control the stereospecificity of polybutadiene with a particular transition metal ion are not very clearly understood at the present time.

IV. Polymerization of Alkynes

Transition metal catalysts employed for the polymerization of alkynes are listed in Table V [242,274–279]. Unlike the polymerization of dienes, alkynes generally yield a multitude of polymers of complex structure along with cyclic aromatic products. Ziegler catalysts obtained from a 1:1 molar ratio of $Ti(OC_4H_9)_4$ and $Al(C_2H_5)_3$ resulted in the formation of high molecular weight acetone-insoluble polymers of acetylene [274]. At higher molar ratios of aluminum to titanium, both acetone-soluble and acetone-insoluble (nonstereospecific) polymers are obtained in good yield. A catalyst system consisting of tris(2,4-pentanediono)chromium(III) and triethylaluminum behaved in a manner similar to that of the titanium catalyst. The structures of the acetone-soluble and acetone-insoluble polymers of acetylene obtained in this way have not yet been elucidated.

Catalysts composed of 2,4-pentanediono complexes of vanadium(IV), iron(III), chromium(III), and nickel(II) and dimethylglyoximino complexes of iron(III), cobalt(III), and nickel(II) (entries 3–9, Table V) produce a high degree of polymerization of phenylacetylene [242]. The activity of a particular catalyst seems to depend mainly on the nature of the transition metal ion and only to a minor extent on the type of ligand attached. Three types of polymers, methanol-soluble but benzene-insoluble, benzene-soluble, and benzene- and methanol-insoluble, have been obtained in each case, the relative amounts varying with the catalyst. The average molecular weight of the benzene-soluble, methanol-insoluble polymer fraction was less than 10^4. The benzene-insoluble polymer was a red bulky solid of unknown molecular weight. The use of bis(dipyridyl)diethyliron(II), diethylnickel(II) complexes (Table V, entries 11,12) in hydrocarbon solvents at low temperatures produces benzene and methanol-insoluble linear high polymers. The linear polymer thus obtained was assigned a structure with repeating —CH=CH— units.

Polymerization of a wide variety of alkynes with bis(triphenylphosphine)-nickel dicarbonyl, **237**, as catalyst has been studied in inert hydrocarbon solvents at reflux temperatures [275] (entry 10, Table V). The formation of aromatic cyclization products from alkynes monosubstituted with alkyl, aryl, vinyl, hydroxymethyl, carbethoxy, acyl, and alkoxy groups has been reported. Monosubstituted alkynes with t-butyl, cyclohexyl, 1-hydroxyethyl, 1-hydroxyisopropyl, 1-hydroxycyclohexyl, and diethylaminomethyl groups yielded mainly linear low molecular weight polymers. Disubstituted acetylenes were generally unreactive.

TABLE V

CATALYSTS FOR THE HOMOGENEOUS POLYMERIZATION OF ALKYNES
BY MULTIPLE INSERTION REACTIONS

No.	Catalyst	Substrate	Temp. (°C)	Polymer	Ref.
1	Ti(OC$_4$H$_9$)$_4$ + Al(C$_2$H$_5$)$_3$	Acetylene	−78, 70	Stereoregular polyacetylenes	274
2	Tris(2,4-pentanediono)-Cr(III) + Al(C$_2$H$_5$)$_3$	Acetylene	−78, 70	Stereoregular polyacetylenes	274
3	Tris(2,4-pentanediono)-V(IV) + Al(C$_2$H$_5$)$_3$	Phenylacetylene	70	High mol. wt. polymer methanol sol., benzene sol., benzene, insol. fractions	242
4	Tris(2,4-pentanediono)-Fe(III) + Al(C$_2$H$_5$)$_3$	Phenylacetylene	70	High mol. wt. polymer methanol sol., benzene sol., benzene, insol. fractions	242
5	Tris(2,4-pentanediono)-Co(III) + Al(C$_2$H$_5$)$_3$	Phenylacetylene	70	High mol. wt. polymer methanol sol., benzene sol., benzene, insol. fractions	242
6	Bis(2,4-pentanediono)-Ni(II) + Al(C$_2$H$_5$)$_3$	Phenylacetylene	70	High mol. wt. polymer methanol sol., benzene sol., benzene insol. fractions	242
7	Bis(dimethylglyoximino)-Fe(III) + Al(C$_2$H$_5$)$_3$	Phenylacetylene	70	High mol. wt. polymer methanol sol., benzene sol., benzene, insol. fractions	242
8	Bis(dimethylglyoximino)-Ni(II) + Al(C$_2$H$_5$)$_3$	Phenylacetylene	70	High mol. wt. polymer methanol sol., benzene sol., benzene, insol. fractions	242
9	Tris(dimethylglyoximino)-Co(III) + Al(C$_2$H$_5$)$_3$	Phenylacetylene	70	High mol. wt. polymer methanol sol., benzene sol., benzene, insol. fractions	242

TABLE V (*continued*)

No.	Catalyst	Substrate	Temp. (°C)	Polymer	Ref.
10	$Ni(CO)_2[(C_6H_5)_3P]_2$	$HC\equiv C-R$ $R = t$-butyl n-pentyl, cyclohexyl $-CH_2CH_2OH$ $-CHOHCH_3$ $-C(OH)(CH_3)_2$ $-CH_2N(C_2H_5)_2$ $-CN$	70	Low mol. wt. polymers with conjugated double bonds	275 276
11	$Fe(C_2H_5)_2(dipy)_2$	Acetylene	$-20, 0$	$-(CH{=}CH)_n-$ $+ C_6H_6$	277
12	$Ni(C_2H_5)_2(dipy)_2$	Acetylene	$-20, 0$	$-(CH{=}CH)_n-$ $+ C_6H_6$	277
13	$RhCl(Ph_3P)_3$	$C_6H_5-C\equiv CH$	50–70	Poly(phenylacetylene)	278
14	$IrH(CO)_2(PPh_3)_2$	$C_6H_5-C\equiv CH$	100–150	Poly(phenylacetylene)	279, 280

The influence of catalyst composition, temperature, and solvent on the rate of polymerization of 1-heptyne has been studied in considerable detail [276]. The activities of substituted triphenylphosphine nickel(0) complexes decrease with substituent in the order $-CH{=}CHCH_3 > -C_6H_5 > -H > -OC_2H_5 > n$-alkyl $\gg -OC_6H_4Cl$. The strong influence of substituents suggests that the triphenylphosphine ligand is retained in the metal complex catalyst during the rate-determining reaction step. Inhibition of the rate of the reaction by carbon monoxide indicates that carbon monoxide undergoes equilibrium dissociation from the nickel complex in the formation of the active catalytic intermediate. No perceivable reaction takes place below 75°C, and direct correspondence between increasing reaction rate and increasing polarity of the solvent was observed with the solvents cyclohexene, benzene, acetonitrile, and methanol. A deuterium isotope effect of ($k_h/k_d = 2.6 \pm 0.4$) has been observed for the polymerization reaction.

1-Heptyne was selected as a substrate for these studies because of its low reactivity at 80°C and exclusive formation of linear dimers and trimers, thus providing a relatively simple system for kinetic analysis. The monomer absorption at 470 nm is of further help in following the rate of the polymerization reaction. The dimer has an absorption maximum at 227 nm and the trimer at 270 nm.

The relative rates of dimer and trimer formation were found to be constant throughout the course of the reaction, the ratio of dimer and trimer remaining at a constant value of 0.7. Addition of free dimer to the reaction mixture did not alter this ratio. It was therefore concluded that free dimer, or a complexed dimer in equilibrium with the free dimer, does not take part in the formation of the trimer. The reaction mechanism involving formulas **237–248** was proposed for the polymerization of alkynes with this catalyst system.

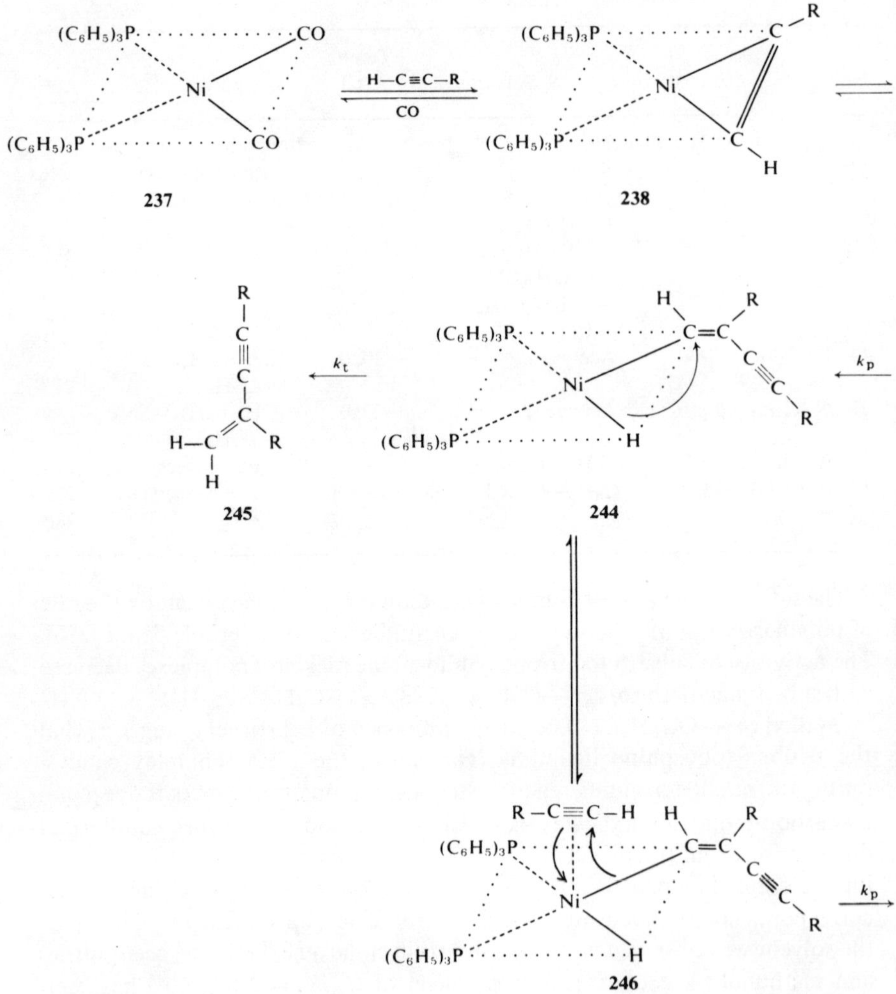

An alkyne–nickel complex **238** is formed in a preequilibrium step by the displacement of 2 moles of carbon monoxide from **237**. Complex **238** then rearranges to the σ-acetylide complex **239**, which has been suggested as the active intermediate for the polymerization of the alkyne. Complex **239** is thus analogous to the metal–alkyl complex formed as an intermediate in the polymerization of alkenes. In subsequent steps a molecule of alkyne is first

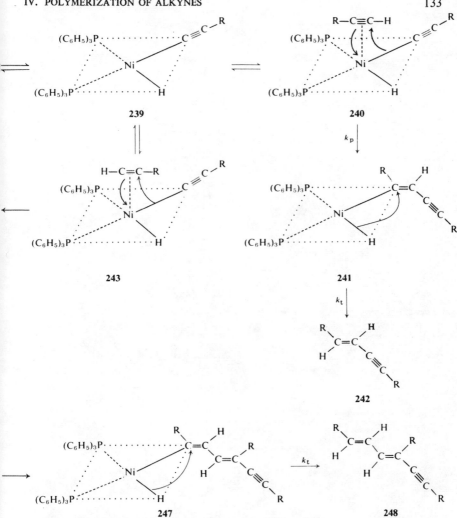

considered to become π-bonded to the metal along an axial bond to form two complexes, **240** and **243**, differing only in the orientation of the π-bonded alkyne. Ligand insertion reactions then take place in complexes **240** and **243** to form σ-bonded vinylacetylene complexes **241** and **244**, respectively. Ligand displacement reaction of the vinylacetylene complexes **241** and **244** by a molecule of alkyne, with the indicated hydride shift, results in the liberation of the free dimers **242** and **245** and metal complex species that can reenter the catalytic cycle at the preequilibrium stage involving **237** ⇌ **238** ⇌ **239**. The suggestion that the intramolecular hydride transfer occurs in the rate-determining step is supported by a large deuterium isotope effect.

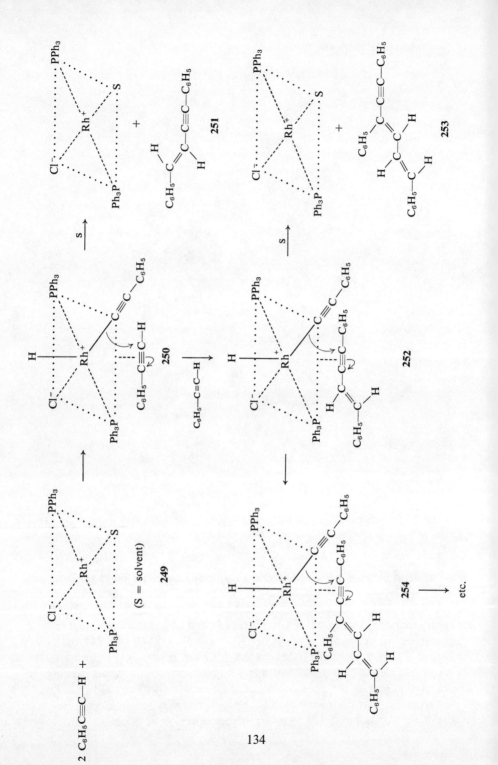

134

The σ-vinylacetylene complexes **241** and **244** may also react with an incoming molecule of acetylene along an axial bond to form intermediate butadienylacetylene σ-bonded complexes such as **247** which can then react further with an alkyne molecule to liberate the trimer. The chain can continue to build up in a manner analogous to these reaction steps to form tetramers and pentamers. Termination and propagation steps take place simultaneously at each stage, resulting in a constant ratio of dimer to trimer to tetramer in the reaction product. The propagation steps k_p have been proposed to be faster than the termination steps, k_t.

The structures of the various dimers and trimers formed in the polymerization reaction of 1-heptyne have been determined by hydrogenation to saturated hydrocarbons and comparison of the saturated products with authentic samples. Further kinetic measurements and structure determinations are required to elucidate the precise nature of insertion and propagation steps in alkyne polymerization.

A reaction similar to the bis(triphenylphosphine)nickel dicarbonyl-catalyzed polymerization of alkynes is the recently reported [278] polymerization of phenylacetylene at 50°–70°C in a bulk monomer medium with bis(triphenylphosphine)chlororhodium(I) as the catalyst. It has been suggested that chain propagation probably proceeds by the addition of phenylacetylide and hydride moieties coordinated to rhodium to a phenylacetylene molecule π complexed to rhodium(III) [278], in accordance with the reaction sequence of formulas **249–254**.

Repeated additions of acetylide and hydride units across the coordinated triple bond in the manner depicted in **250**, **252**, and **254** lead to chain growth. 1,4-Diphenyl-*trans*-but-1-ene-3-yne **251** has been obtained in 50% yield when the reaction is conducted at 0°–20°C with 0.016 mole/liter of a rhenium catalyst of unspecified composition.

Brown *et al.* [279] have reported the polymerization of phenylacetylene to mainly a linear polymer by heating the alkyne with $IrH(CO)_2(PPh_3)_2$ under reflux for 24 hours. Along with the linear polyphenylacetylene about 6% of 1,3,5- and 1,2,4-triphenylbenzene was obtained.

V. Oligomerization of Ethylene and α-Olefins

Oligomerization of ethylene and α-olefins, and co-oligomerization of alkenes are best carried out at low temperatures (e.g., $-50°C$, $-100°C$) with Ziegler catalysts composed of monomethyltitanium chloride and alkylaluminum chlorides in chlorinated hydrocarbon solvents. The products of oligomerization of ethylene consist of straight-chain α-olefins and 2-ethyl-1-alkenes. Since α-olefins react more slowly with the catalyst than does ethylene, a mixture of low molecular weight oligomers containing 60–65% of the respective dimers is formed. Introduction of ethylene to a catalyst solution containing an α-olefin results in highly selective β-ethylation of the latter. In all catalytic oligomerization reactions of this type, the system remains homogeneous, with presumably no change in the valence state of titanium.

In order to gain insight into the nature of the active intermediate in the titanium–aluminum catalyst system, varying molar ratios of the two components CH_3TiCl_3 and CH_3AlCl_2 (Ti:Al molar ratio varied from 2 to 0.5), were dissolved in methylene chloride solution at $-80°C$ and treated with ethylene or other olefin [225]. According to any reasonable mechanism, the alkyl group at the active site of the catalyst should react with an olefin molecule, but the alkyl group at the inactive site would be expected to remain undisturbed and may be displaced at a later stage by treatment with an alcohol producing the alkane and destroying the catalyst. At all titanium: aluminum ratios employed, the quantity of methane liberated on treatment with an alcohol after reaction with an olefin was exactly equivalent to the quantity of CH_3AlCl_2 present in the catalyst. It was therefore deduced that the methyl group attached to the aluminum co-catalyst remains intact during oligomerization of the olefin and that the reaction proceeds exclusively through the titanium alkyl complex. The presence of the aluminum compound in the catalyst activates the titanium–carbon bond in a manner that is not yet clear. The oligomerization of 1-pentene by the CH_3TiCl_3–CH_3AlCl_2 catalytic mixture at $-70°C$ is about 300 times faster than that observed when the titanium complex alone was employed. The presence of chlorinated hydrocarbon solvents such as methylene chloride and chloroform enhance the rate of oligomerization of ethylene even at $-70°C$. From the effect of polar solvents on the rate of oligomerization it has been deduced that the active catalyst is either a chloro-bridged binuclear complex, **255**, or an ion pair, **256**.

255   **256**

Chlorinated and aromatic hydrocarbons help to increase the rate of oligomerization by solvating the positive methyldichlorotitanium species. The formation of a solvated catalyst center is further supported by intensification of the color of the catalyst in methylene chloride or toluene at low temperatures. (The catalyst regains its original color at room temperature.) The mechanism proposed [225] for catalyzed oligomerization of ethylene with this titanium–aluminum catalyst is illustrated by the reaction sequence involving formulas **257–261**. An alkyltitanium complex **257** (aluminum and chloride ions are not

$$
\underset{257}{\overset{\displaystyle \text{Ti}^{3+}\!---\!}{\underset{\displaystyle |}{\text{C}_2\text{H}_5}}} + \text{C}_2\text{H}_4 \longrightarrow \underset{258}{\overset{\displaystyle \text{Ti}^{3+}\!---\!\overset{\text{CH}_2}{\underset{\text{CH}_2}{\|}}}{\underset{\displaystyle |}{\text{C}_2\text{H}_5}}} \xrightarrow{\text{C}_2\text{H}_4} \underset{259}{\overset{\displaystyle \text{Ti}^{3+}\!---\!\overset{\text{CH}_2}{\underset{\text{CH}_2}{\|}}}{\underset{\displaystyle |}{\underset{\displaystyle \text{CH}_2\!-\!\text{CH}_2\!-\!\text{C}_2\text{H}_5}{}}}}
$$

$$
257 + \text{1-butene} \xleftarrow{\text{C}_2\text{H}_4} \underset{260}{\overset{\displaystyle \text{Ti}^{3+}\!-\!\text{C}_2\text{H}_5}{\underset{\displaystyle |}{\underset{\displaystyle \text{C}_2\text{H}_5}{\text{H}_2\text{C}\!=\!\text{CH}}}}} \qquad \underset{261}{\overset{\displaystyle \text{Ti}^{3+}\!---\!\overset{\text{CH}_2}{\underset{\text{CH}_2}{\|}}}{\underset{\displaystyle |}{\underset{\displaystyle (\text{CH}_2)_4\!-\!\text{C}_2\text{H}_5}{}}}}
$$

shown for clarity) combines with ethylene to form a π complex **258**, as indicated by an intensification of the color of a methylene chloride solution of the catalyst from yellow to dark red. An olefin insertion reaction then takes place by a mechanism similar to that proposed by Cossee [247] and a new butyltitanium π-ethylene complex, **259**, is obtained. The latter can then undergo either further olefin insertion to form a hexyltitanium–ethylene complex, **261**, or can take part in a chain-transfer reaction by an intramolecular shift of hydride ion from the β-carbon atom of the butyl group in **259** to a π-bonded ethylene, resulting in the formation of a new ethyltitanium π-butene complex, **260**. The chain-transfer reaction **259** → **260** is reversible, and some further chain growth can occur. The earlier steps of this oligomerization mechanism (**257–259**) are identical to the corresponding part of the Ziegler polymerization sequence (**220–223**), with chain transfer to the monomer **259–260** making oligomerization possible.

The ethyltitanium π-butene complex **260** can take part in two possible reactions, indicated by sequences (223) and (224). Olefin insertion by β-ethylation of the olefin [Markownikoff addition, Eq. (223)] gives the 1-alkyl-π-ethylene complex **262**, while α-addition to ethylene forms the 3-alkyltitanium π-ethylene complex **264**. Chain-transfer reactions followed by displacement result in the formation of 2-ethyl-1-butene, 2-hexene, and 3-hexene.

The insertion reactions of the higher olefins are faster than the displacement of the olefin by ethylene. This results in the formation of a high ratio of branched α-olefins and internal olefins to straight-chain oligomers. Isomerization of the internal olefins 2-hexene and 3-hexene takes place presumably by hydride transfer to ethylene in complex **264** from neighboring carbon atoms,

$$\cdots Ti^{3+}\!\!-\!C_2H_5 \;\xrightarrow{C_2H_4}\; \cdots Ti^{3+}\!\cdots\!\Big\|\begin{array}{c}CH_2\\CH_2\end{array} \;\longrightarrow\; \cdots Ti^{3+}\!\!-\!C_2H_5 \;\longrightarrow\; \text{2-ethyl-1-butene}$$

$$\underset{\mathbf{260}}{H_2C\!\!=\!\!CHC_2H_5} \qquad \underset{\mathbf{262}}{\underset{\displaystyle C_2H_5}{H_2C\!-\!CH\!-\!C_2H_5}} \qquad \underset{\mathbf{263}}{H_2C\!\!=\!\!C(C_2H_5)_2}$$

$$(223)$$

$$\cdots Ti^{3+}\!\!-\!C_2H_5 \;\xrightarrow{C_2H_4}\; \cdots Ti^{3+}\!\cdots\!\Big\|\begin{array}{c}CH_2\\CH_2\end{array} \;\longrightarrow\; \cdots Ti^{3+}\!\!-\!C_2H_5 \;\xrightarrow{C_2H_4}\; \text{3-hexene}$$

$$\underset{\mathbf{260}}{C_2H_5HC\!\!=\!\!CH_2} \qquad \underset{\mathbf{264}}{C_2H_5CH\!-\!CH_2C_2H_5} \qquad \underset{\mathbf{265}}{\underset{\displaystyle H_5C_2 \quad C_2H_5}{HC\!\!=\!\!CH}}$$

$$\cdots Ti^{3+}\!\!-\!C_2H_5 \;\xrightarrow{C_2H_4}\; \text{2-hexene}$$

$$\underset{\mathbf{266}}{CH_3\!-\!CH\!\!=\!\!CH\!-\!CH_2C_2H_5}$$

$$(224)$$

resulting in the formation of 2-hexene and 3-hexene π complexes **266** and **265**, respectively. Further oligomerization to tetramers, pentamers, and so on, can take place by ligand insertion and chain-transfer reactions similar to those illustrated in Eqs. (223) and (224).

The mechanism of oligomerization shown for ethylene in the reactions discussed above may also apply to isomerization of other alkenes. In such cases, double-bond migration does not generally proceed much beyond two positions, because of facile displacement of alkenes from the alkyltitanium–π-olefin complexes. The process of β- or α-ethylation of higher alkenes (co-oligomerization) is similar to the corresponding reactions of ethylene, with the added difference that migration of the double bonds does not occur under the condition of co-oligomerization.

Simple dimerization of ethylene with no side products may be carried out in an atmosphere of ethylene at temperatures as high as 0°C with Ziegler catalysts consisting of organoaluminum compounds such as triethylaluminum, diisobutylaluminum hydride, ethoxydiethylaluminum, and sodium tetraethylaluminate, together with 2,4-pentanediono complexes of cobalt(II) or cobalt(III) in benzene, n-heptane, diethyl ether, or tetrahydrofuran solution [280]. A reduced cobalt complex is assumed to be the effective catalyst for the dimerization of ethylene. The highest yield (36.2%) of n-butene for this reaction was observed with a molar ratio of aluminum to cobalt complex of 4:1. Larger proportions of aluminum complexes reduce the catalytic activity by conversion of the catalyst to unreactive forms. The n-butenes obtained in the reaction consist of 95% 2-butene and 5% 1-butene. The high percentage of 2-butene in the reaction mixture is due to the fast isomerization of 1-butene by the catalytic system.

The rhodium(III)-catalyzed dimerization of ethylene in protic solvents was reported by Cramer [208,209] and by Alderson et al. [214]. Dimerization of ethylene at 1000 atm in an alcoholic solution of rhodium(III) chloride resulted in the formation of a mixture of 1-butene and 2-butene, with no higher molecular weight products. The conversion and isomer distribution of the butene mixtures depend on temperature, reaction time, and quantity of the catalyst. At 30°C 1-butene constitutes 38% of the isomer mixture, whereas at 45°C, the composition of the mixture corresponds to 2% 1-butene, 78% trans-2-butene, and 20% cis-2-butene. The higher yield of 1-butene at low temperatures and a shorter reaction time is considered due to its formation as the primary oligomerization product. Rhodium(III) gave a mixture of butenes at 50°C [214]. At 150°C, however, higher oligomers of ethylene are obtained along with the butenes [214].

Cramer [208,209] has investigated kinetics of dimerization of ethylene by rhodium(III) and has presented the mechanism indicated by Eqs. (225)–(230). Isolation of most of the intermediates indicated in these reactions provided evidence for the proposed mechanism.

$$RhCl_3 + C_2H_5OH + 2\ C_2H_4 \longrightarrow \underset{267}{Cl_2Rh(C_2H_4)_2^-} + Cl^- + 2\ H^+ + CH_3CHO \tag{225}$$

$$\underset{267}{Cl_2Rh(C_2H_4)_2^-} + H^+ + Cl^- \rightleftharpoons \underset{268}{C_2H_5RhCl_3(C_2H_4)S^-} \tag{226}$$

$$\underset{268}{C_2H_5RhCl_3(C_2H_4)S^-} \rightleftharpoons C_2H_4 + \underset{269}{C_2H_5RhCl_3S^-} \tag{227}$$

$$\underset{268}{C_2H_5RhCl_3(C_2H_4)S^-} \xrightarrow[\text{slow}]{k} \underset{270}{CH_3CH_2CH_2CH_2RhCl_3S^-} \tag{228}$$

$$\underset{270}{CH_3CH_2CH_2CH_2RhCl_3S^-} \xrightarrow{-H^+,\ -Cl^-} \underset{271}{(CH_3CH_2CH{=}CH_2)Cl_2RhS^-} \tag{229}$$

$$\underset{271}{(CH_3CH_2CH{=}CH_2)Cl_2RhS^-} \xrightarrow{C_2H_4} \underset{272}{[Cl_2Rh(C_2H_4)_2]^-} + C_4H_8 \tag{230}$$

$$\text{(where S = solvent)}$$

The complex anion **267** is obtained by the interaction of rhodium trichloride with ethylene. Its formation has been confirmed by spectrophotometric evidence for the formation of the same anion by the reaction of $[ClRh(C_2H_4)_2]_2$ with chloride ion in alcoholic solution according to Eq. (231).

$$[ClRh(C_2H_4)_2]_2 + 2\ Cl^- \longrightarrow 2\ [(C_2H_4)_2RhCl_2]^- \tag{231}$$

The bis(ethylene) complex of monovalent rhodium **267** is rapidly converted by reaction with hydrochloric acid into an octahedral ethylrhodium complex, **268**, accompanied by a change in color from yellow to orange-red (absorption at 600 mμ). Coordination of three chloride ions to the metal in **268** was substantiated by potentiometric measurements and by the observation of a linear increase in the rate of dimerization of ethylene with chloride concentration only above a molar chloride to rhodium ratio of 3:1. This trichlororhodium ion, **268**, has been formulated as a six-coordinate complex on the basis of its elemental composition and electronic spectrum. The composition indicated was also supported by NMR measurements indicating the presence of both ethylene and ethyl groups in the coordination sphere of rhodium.

As indicated in Eq. (227), complex **268** dissociates reversibly to ethylene and a new rhodium(III) complex **269**. Evidence for **269** was found by its NMR spectrum in perdeuteromethanol, which showed peaks characteristic of a coordinated ethyl group. An alternative to the reaction shown above for the formation of **268** is the formation of an intermediate rhodium hydride with subsequent addition of hydride to one of the ethylene molecules, as indicated by the reaction sequence (232), (232a).

$$[Cl_2Rh(C_2H_4)_2]^- + HCl \longrightarrow [Cl_3Rh(C_2H_4)_2H]^- \tag{232}$$

$$[Cl_3Rh(C_2H_4)_2H]^- \xrightarrow{\text{solvent}} [(C_2H_5)RhCl_3(C_2H_4)S]^- \tag{232a}$$

The presence of chloride ions in the coordination sphere of rhodium(III) seems to be essential for the dimerization reaction of ethylene since catalytic activity is lost if hydrochloric acid is replaced by sulfuric acid, acetic acid, nitric acid, or trifluoroacetic acid.

The proposed rate-determining step, (228), in the dimerization reaction is olefin insertion in the ethyl–rhodium σ bond with the formation of a σ-butylrhodium complex, **270**. The olefin insertion reaction has been substantiated by a change in the NMR spectra of the ethyl–rhodium complex at the polymerization temperature of ethylene (10°–50°C), whereas no change was observed below this temperature range. The rate constant k for the dimerization of ethylene has been reported [208] as $7.4 \times 10^8 \, e^{-17.2/RT}$ ($\Delta H^{\ddagger} = 16.16$ kcal/mole, $\Delta G^{\ddagger} = 22.7$ kcal/mole, $\Delta S^{\ddagger} = -20.1$ eu). The final step, (229), in the dimerization of ethylene is the intramolecular formation of π-bonded 1-butene by the decomposition of the butylrhodium anion **270**. An alternative bimolecular chain-transfer mechanism with concerted intermolecular hydride transfer has been ruled out on the basis of the observed change in the oxidation state of rhodium(I) to rhodium(III) by reaction of **271** with hydrogen chloride and the subsequent reduction of rhodium(III) back to rhodium(I) complex **272** accompanied by isomerization of 1-butene to 2-butene, as indicated by Eq. (233). Final displacement [Eq. (230)] of 1-butene or 2-butene from the butene–Rh(I) chloride complex by an incoming ethylene molecule regenerates the original form of the catalyst. The isomerization and displacement steps are faster than the olefin insertion reaction. Thus 1-butene or 2-butene could be displaced from the butene–Rh(I) chloride complex, before the incoming ethylene is coordinated and inserted. The displacement reaction is also aided by the fact that the affinity of ethylene for rhodium(I) is higher than the affinities of the butenes.

$$\underset{\displaystyle \overset{|}{\underset{}{CH_3CH_2CHCH_3}}}{\overset{Rh^{2+},\ Cl^-}{}} \longrightarrow \underset{\displaystyle \overset{Rh^+}{CH_3CH \overset{\vdots}{=} CHCH_3}}{} + HCl \qquad (233)$$

Oligomerization of methyl acrylate is catalyzed by rhodium(III) complexes or rhodium(III) chlorides in methanol [214]. A high yield of the dimer, dimethyl 2-hexenedioate, has been obtained by heating a solution of methyl acrylate in methanol at 140°C in the presence of rhodium trichloride [Eq. (234)].

$$2\ CH_2{=}CHCOOCH_3 \longrightarrow CH_3OOCCH{=}CHCH_2CH_2COOCH_3 \qquad (234)$$

VI. Oligomerization of Dienes and Alkynes

A. Linear Oligomerization and Co-oligomerization of Dienes

Transition metal ion catalysts employed for the linear oligomerization and co-oligomerization of dienes are tabulated in Table VI [81,211,212,214,282–296]. Catalytic systems effective for oligomerization of dienes are composed mainly of low-valent cobalt, nickel, iron, rhodium, titanium, and palladium complexes. The most widely used catalyst by far is zero-valent cobalt (entries 1–8, Table VI).

Treatment of butadiene in a hydrocarbon solution with a cobaltous chloride–trialkylaluminum catalyst resulted [282] in the formation of two dimers, 85% 3-methyl-1,4,6-heptatriene, **273**, and 15% 1,3,6-octatriene, **274** (entry 1, Table VI). These dimers are formed from 1,3-butadiene by 1,2- and 1,4- addition, as indicated by Eq. (235). The products were characterized by

$$CH_2{=}CH{-}CH{=}CH_2 \begin{cases} \overset{\displaystyle 1,2\text{-addition}}{\nearrow} & CH_2{=}CH{-}\underset{\displaystyle \underset{CH_3}{|}}{CH}{-}CH{=}CH{-}CH{=}CH_2 \\ & \textbf{273} \\ \underset{\displaystyle 1,4\text{-addition}}{\searrow} & \\ & CH_2{=}CH{-}CH{=}CH{-}CH_2{-}CH{=}CH{-}CH_3 \\ & \textbf{274} \end{cases} \qquad (235)$$

hydrogenation, infrared, and NMR spectra. The highest yield (90%) of the dimers is obtained when triethylaluminum is employed. Polymers of butadiene are formed only in traces.

TABLE VI

CATALYTIC OLIGOMERIZATION AND CO-OLIGOMERIZATION OF DIENES

No.	Catalyst	Solvent	Substrate	Temp. (°C)	Products	Ref.
1	CoCl₂ + Al(C₂H₅)₃	Hydrocarbons	Butadiene	—	1,3,6-Octatriene + 3-methyl-1,4,6-heptatriene (15%, 85%)	282
2	Co₂(CO)₈ + Al(C₂H₅)₃	Hydrocarbons	Butadiene	40	trans-3-Methyl-1,4,6-heptatriene (90%)	283
3	CoCl₂ + Al(C₂H₅)₃	Hydrocarbons or ethers	Butadiene	30–100	1,3,6-Octatriene + 3-methyl-1,4,6-heptatriene (10%, 90%)	284
4	CoCl₂ + Al(C₂H₅)₃	Hydrocarbons or ethers	Butadiene + ethylene	80	trans-1,3-Hexadiene	285
5	CoCl₂ + Al(C₂H₅)₃	Hydrocarbons or ethers	Butadiene + acrylic esters	40–60	Heptadienoic esters (4,6-heptadiene-1-carboxylic esters)	285
6	CoCl₂ + Al(C₂H₅)₃	Hydrocarbons or ethers	Isoprene + styrene	40–60	1-Phenyl-4,8-dimethyldecatriene	286
7	Tris(2,4-pentanediono)-Co(III) + Al(C₂H₅)₃	Hydrocarbons or ethers	Butadiene	20	3-Methyl-1,4,6-heptatriene	287
8	Tris(2,4-pentanediono)-Co(III) + Al(C₂H₅)₃	Hydrocarbons or ethers	Butadiene + ethylene	20	n-hexa-1,3-diene (70%)	288
9	Tris(2,4-pentanediono)-Fe(III) + Al(C₂H₅)₃ (1:1)	Hydrocarbons or ethers	Butadiene	40	n-Dodeca-1,3,6,10-tetraene + dimers	286
10	Tris(2,4-pentanediono)-Fe(III) + Al(C₂H₅)₃	Toluene	Butadiene + ethylene	30	cis-1,4-Hexadiene (35%) 2,4-hexadiene, 1,3-hexadiene	81
11	Tris(2,4-pentanediono)-Fe(III) + Al(C₂H₅)₃	Toluene	1,3-Pentadiene + ethylene	50	3-Methyl-cis-1,4-hexadiene + cis-1,3-heptadiene	81

12	Tris(2,4-pentanediono)-Fe(III) + Al(C$_2$H$_5$)$_3$	Toluene	Isoprene + ethylene	20	4-Methyl-1,4-hexadiene + 5-methyl-1,4-hexadiene	81
13	Tris(2,4-pentanediono)-Fe(III) + Al(C$_2$H$_5$)$_3$	Toluene	2,3-Dimethyl-1,3-butadiene + ethylene	—	4,5-Dimethyl-1,4-hexadiene	81
14	Acrylonitrile bis(tri-o-tolylphosphite)-Ni(0)	Alcohol	Butadiene	30	n-Octatriene + 3-methyl-1,4,6-heptatriene	285
15	(R$_3$P)$_m$Ni(CO)$_n$ ($m + n = 4$)	Alcohol	Butadiene	—	Octadienes, octatrienes + trimers	289
16	RhCl$_3$ + pot. acetate	Protic	Butadiene	100	70% dimers, 30% high mol. wt. products	214
17	RhCl$_3$	Protic	Butadiene + ethylene	50	1,4-Hexadiene + 2,4-hexadiene	214
18	RhCl$_3$	Protic	1,3-Pentadiene + ethylene	50	3-Methyl-1,4-hexadiene	214
19	RhCl$_3$	Protic	Isoprene + ethylene	50	4-Methyl-1,4-hexadiene	214
20	RhCl$_3$	Protic	Butadiene + propylene	50	2-Methyl-1,4-hexadiene	214
21	RhCl$_3$	Protic	Isoprene + propylene	50	2,4-Dimethyl-1,4-hexadiene	214
22	TiCl$_4$ + (CH$_2$=CH)MgBr	THF	Butadiene	25	4-Vinylcyclohexene + 2,6-octadiene	290
23	(PPh$_3$)$_2$Pd (maleic anhydride)	C$_6$H$_6$ THF	Butadiene	100–120	3-Methyl-1,5-heptadiene + 3-methyl-1,4,6-heptatriene + 1,3,7-octatriene	291
24	(PPh$_3$)$_2$Pd (maleic anhydride)	CH$_3$OH C$_2$H$_5$OH	Butadiene	40–100	8-Alkoxy-1,6-octadiene + 3-alkoxy-1,7-octadiene + 1,3,7-octatriene	291

TABLE VI (*continued*)

No.	Catalyst	Solvent	Substrate	Temp. (°C)	Products	Ref.
25	$(PPh_3)_2Pd$ (maleic anhydride)	C_6H_6	Butadiene + ethylene	100–120	1,4-Hexadiene	291
26	$RuCl_3 \cdot 3H_2O$	Ethanol	CH_2=CH—CN	130	*cis*- and *trans*-1,4-Dicyano-1-butene	292 293 294 295
27	$(PPh_3)_2Pd$ (maleic anhydride)	*n*-butylamine	Butadiene	—	1-(*n*-Butylamino)-3,7-octadiene	296
28	$(PPh_3)_2Pd$ (maleic anhydride)	BH BH = morpholine, piperidine diisopropyl- amine	Butadiene	—	1-B-Substituted-2,7-octadiene	296
29	$CoCl_2$(diphos) + $AlEt_3$	$C_2H_4Cl_2$	Butadiene + ethylene	80–110	*cis*-1,4-Hexadiene	211
30	$HNi[P(OEt)_3]_4^+$	Methanol	Butadiene + ethylene	50–60	*trans*-1,4-Hexadiene 70% + *cis*-1,4-hexadiene + 3-methyl-1,4-pentadiene 30%	212

The use of dicobalt octacarbonyl and triethylaluminum as catalyst resulted in the dimerization of butadiene to give a 90% yield of *trans*-3-methyl-1,4,6-heptatriene (entry 2, Table VI). The cobalt carbonyl catalyst system contained a large excess of aluminum alkyl [283] [80 mmole of $Co_2(CO)_8$ per 2.4 mole of $Al(C_2H_5)_3$]. In the absence of the monomer, metallic cobalt is precipitated. Infrared carbon monoxide stretching bands were completely absent from the catalyst system both in the presence and in the absence of the monomer. It was therefore proposed that the active oligomerization catalyst is zero-valent cobalt with a d^9 configuration, loosely coordinated to butadiene. Use of the same catalysts in solvent mixtures containing methylene chloride produced solid polybutadiene polymers with high *cis*-1,4-character with small quantities of 4-vinylcyclohexene, *n*-octatriene, and 1,5-cyclooctadiene as by-products [283].

Zero-valent cobalt has been proposed by Wittenberg [284] and by Muller *et al.* [285] as the catalyst in the oligomerization and co-oligomerization of dienes. Catalytic solutions containing reduced cobalt species have been obtained by reduction of cobalt(II) compounds with aluminum trialkyl in organic solvents [284,285]. These solutions catalyze oligomerization of butadiene to open-chain dimers, trimers, and tetramers (entries 3–6, Table VI). The dimer fraction consisted of 90% 3-methyl-1,4,6-heptatriene and 10% 1,3,6-octatriene. If the reaction is allowed to proceed further the yield of the dimer gradually decreases with time with a concomitant increase in the yield of higher oligomers.

Oligomerization of butadiene with catalysts consisting of tris(2,4-pentanediono)cobalt(III) and Fe(III) complexes in the presence of aluminum trialkyl (entries 7–13, Table VI) has been reported by Saito *et al.* [287,288], by Hidai *et al.* [286] and by Hatta [81]. The order of mixing of the components has a considerable effect on the yield of the oligomers. When triethylaluminum is added first to the cobalt(III) or the iron(III) complex species with the subsequent addition of butadiene, the yield of oligomers is greatly reduced. It has been proposed that such inactivation of cobalt(III) and iron(III) catalysts by aluminum triethyl is due to the formation of reduced metal ions that are precipitated as solid metal species and as such can no longer act as catalysts in the homogeneous system [287,288]. If the order of addition of the components is reversed, however, the solutions show marked reactivity towards oligomerization of butadiene. It was proposed that the zero-valent cobalt and iron species formed in solution by the reduction of the 2,4-pentanediono complexes are stabilized sufficiently, in the presence of diene, to remain in solution and catalyze oligomerization of butadiene. The cobalt catalyst caused selective dimerization of butadiene to 3-methyl-1,4,6-heptatriene [287] (entry 7, Table VI). The iron catalyst, however, gave a trimer, dodeca-1,3,6, 10-tetraene [286] (entry 9, Table VI). The maximum yield of the trimer was obtained when the molar ratio of iron to aluminum was 1:1. Increase in the ratio to more than 3:1 resulted in extensive polymerization of butadiene as the main reaction.

A catalyst based on zero-valent nickel has been obtained by the interaction of bis(acrylonitrile)nickel(0) with bis(tri-o-tolylphosphite) [285]. In the presence of alcohols the complex dimerized butadiene to isomeric octatrienes, consisting mainly of trans-1,3,7-octatriene, together with small amounts of 3-methyl-1,4,6-heptatriene (entry 14, Table VI). In the absence of alcohols, the catalyst promotes cyclodimerization of butadiene to form 1,5-cyclooctadiene in 90% yield. Mixed trialkyl- or triarylphosphine nickel carbonyl complexes in alcohol solution also yield octatrienes, octadienes, and trimers [289] (entry 15, Table VI). In the absence of alcohols, the nickel(0) catalyst dimerizes and trimerizes butadiene to cyclic products.

Dimerization of butadiene with rhodium trichloride and potassium acetate in alcohol solution at 100°C resulted in 54% conversion to a mixture containing 70% of dimers and 30% of higher molecular weight products [214]. The dimers consisted mainly of 2,4,6-octatrienes (entry 16, Table VI).

Reaction of a mixture of ethylene and butadiene at 50°C in the presence of a rhodium(III) catalyst was found to produce a mixture of 1,4-hexadiene and 2,4-hexadiene [214] (entry 17, Table VI). It has been suggested that the 1,4-diene is the primary product of the co-oligomerization reaction and is subsequently isomerized to the 2,4-diene [214]. Alkyl derivatives of 1,4-hexadiene have been obtained by co-oligomerization of isoprene and 1,3-pentadiene with ethylene, and of isoprene with 1-propene (entries 17–21, Table VI). The formation of 1,4-hexadiene may be visualized as the 1,4-addition of the vinyl group and hydrogen to one of the double bonds of butadiene [Eq. (236)].

$$CH_2{=}CH_2 + CH_2{=}CH{-}CH{=}CH_2 \xrightarrow{\text{Rh(III)}} CH_2{=}CH{-}CH_2{-}CH{=}CH{-}CH_3$$

$$(236)$$

Since butenes are not formed during the co-oligomerization process, it has been proposed by Cramer [209,296] that ethylene and butadiene form a mixed-ligand π complex with rhodium(III) of higher stability than the 2:1 π complexes of ethylene and of butadiene. In the proposed mechanism, a coordinated partially negative π-allyl butenyl group (formed by hydride addition to a coordinated butadiene) reacts with a coordinated ethylene to form a chelated σ,π-rhodium(III) complex, 276. Addition of a second molecule of butadiene results in the formation of the mixed-ligand complex 277 that rearranges to the π-1,4-hexadiene complex 278. Replacement of 1,4-hexadiene by ethylene regenerates complex 275. At low temperatures (below 30°C) ethylene insertion in the coordinated butenyl group (reaction 275 → 276) is rate-determining while at 30°–50°C, rearrangement of the complex (reaction 277 → 278) is the proposed rate-determining step [296].

The dimerization of butadiene with bis(triphenylphosphine)(maleic anhydride)palladium(II) at 100°–120°C in aprotic solvents such as benzene, tetrahydrofuran, and acetone resulted in the selective formation of 1,3,7-octatriene in good yield [291] (entries 23–25, Table VI). At 115°C, the yields

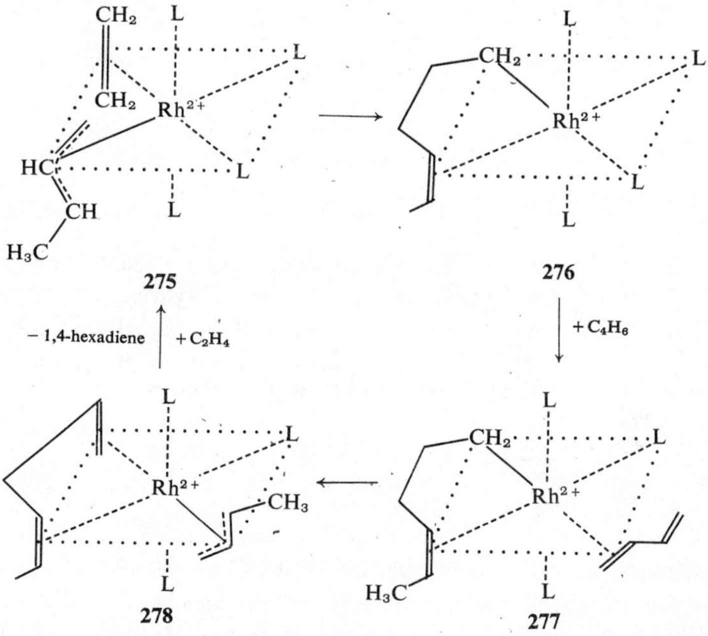

of 1,3,7-octatriene in tetrahydrofuran and acetone are 82% and 86%, respectively, while at 120°C in benzene the yield drops to 64%. Above 120°C rapid decomposition of the palladium(II) catalyst and loss in the catalytic activity occurred. In methanol, ethanol, and isopropanol, however, butadiene is converted into 1-alkoxy-2,7-octadiene, **279**, and 3-alkoxy-1,7-octadiene, **280**, and/or 1,3,7-octatriene, **281**, depending on the nature of the alcohol (entry 24, Table VI). The overall reaction in benzene or in an alcohol are illustrated [291] by reactions (237) and (238).

$$2\ CH_2{=}CH{-}CH{=}CH_2 \xrightarrow[\text{ROH}]{\text{Pd(II)}} R{-}O{-}CH_2{-}CH{=}CH{-}(CH_2)_3{-}CH{=}CH_2 \quad (\textit{trans})$$
$$\mathbf{279} \qquad\qquad (237)$$
$$+$$
$$H_2C{=}CH{-}CH(OR){-}(CH_2)_3{-}CH{=}CH_2$$
$$\mathbf{280}$$

$$2\ CH_2{=}CH{-}CH{=}CH_2 \xrightarrow[\text{C}_6\text{H}_6]{\text{Pd(II)}} CH_2{=}CH{-}CH{=}CH{-}CH_2{-}CH_2{-}CH{=}CH_2$$
$$\mathbf{281} \qquad\qquad (238)$$

Adducts of the composition $R_2NC_8H_{13}$ and $RNHC_8H_{13}$ have been obtained [297] when the dimerization of butadiene is conducted in secondary amines (morpholine, piperidine, diisopropylamine) and in primary amines (*n*-butylamine), respectively, (entries 27, 28, Table VI), as indicated by reactions (239) and (240).

$$2\ CH_2{=}CH{-}CH{=}CH_2 + BH \xrightarrow{Pd(II)} B{-}CH_2{-}CH{=}CH{-}(CH_2)_3{-}CH{=}CH_2$$
$$\mathbf{282} \qquad\qquad (239)$$

$$2\ CH_2{=}CH{-}CH{=}CH_2 + RNH_2 \xrightarrow{Pd(II)} RNH{-}CH_2{-}CH{=}CH{-}(CH_2)_3{-}CH{=}CH_2$$
$$\mathbf{283} \qquad\qquad (240)$$

(BH = piperidine, morpholine, diisopropylamine)

End-to-end dimerization of acrylonitrile has been effected with a ruthenium trichloride trihydrate catalyst under 3.7 atm of hydrogen in ethanol at 130°C [292–295]. A mixture of *cis*- and *trans*-1,4-dicyano-1-butene has been obtained in 44% overall yield, along with a 51% yield of propionitrile (entry 26, Table VI). With rhodium trichloride, iridium tetrachloride, and niobium

$$CH_2{=}CHCN \xrightarrow[H_2\ (3.7\ atm)]{RuCl_3 \cdot 3\ H_2O} CH_3{-}CH_2{-}CN\ (51\%) + NC{-}CH{=}CH{-}CH_2{-}CH_2{-}CN$$
$$(\textit{cis and trans})\ 44\%$$
$$(241)$$

pentachloride as catalysts, only propionitrile was obtained.

Linear co-oligomerization of butadiene and ethylene in hydrocarbon solvents is catalyzed by zero-valent cobalt at 80°C. The catalyst solution prepared from cobaltous chloride and triethylaluminum produced very pure *trans*-1,3-hexadiene. The selective formation of this isomer was explained [285] as occurring by a hydrogen shift from butadiene to ethylene with a subsequent carbon–carbon coupling reaction. The term "dienylation" has

$$CH_2{=}CH{-}CH{=}CH_2 + CH_2{=}CH_2 \xrightarrow{Co(0)} H_2C{=}CH{-}CH{=}CH{-}CH_2{-}CH_3$$
$$(242)$$

been given to this reaction [Eq. (242)] by Wittenberg [284]. Equations (243)–(248) illustrate the mechanism proposed [225,271] for the Co(0)-catalyzed co-oligomerization of ethylene and butadiene.

Alkylation of cobalt and subsequent reduction as indicated in Eqs. (243) and (244) were proposed by Natta *et al.* [271] as the initial steps in the formation of the reactive catalytic intermediate in the polymerization of butadiene with tris(2,4-pentanediono)cobalt(III) and triethylaluminum. The ethyl radical formed from diethylcobalt in reaction (244) rapidly disproportionates to ethylene and ethane [reaction (245)]. Ethylcobalt then forms a π complex,

284, with butadiene. Transfer of the ethyl group to the α-carbon atom of the π-bonded butadiene followed by coordination of a molecule of ethylene yields the hexenyl-π-ethylene complex **285**. Intramolecular transfer [Eq. (247)] of a hydrogen atom to coordinated ethylene in complex **285** produces an ethylcobalt-1,3-hexadiene complex, **286**, from which 1,3-hexadiene is displaced by a molecule of butadiene [Eq. (248)], regenerating the π-butadiene ethylcobalt complex **284**.

$$CoCl_2 + (C_2H_5)_3Al \rightleftharpoons (C_2H_5)_2Co + C_2H_5AlCl_2 \qquad (243)$$

$$(C_2H_5)_2Co \longrightarrow C_2H_5Co + C_2H_5 \cdot \qquad (244)$$

$$2\,C_2H_5 \cdot \longrightarrow C_2H_4 + C_2H_6 \qquad (245)$$

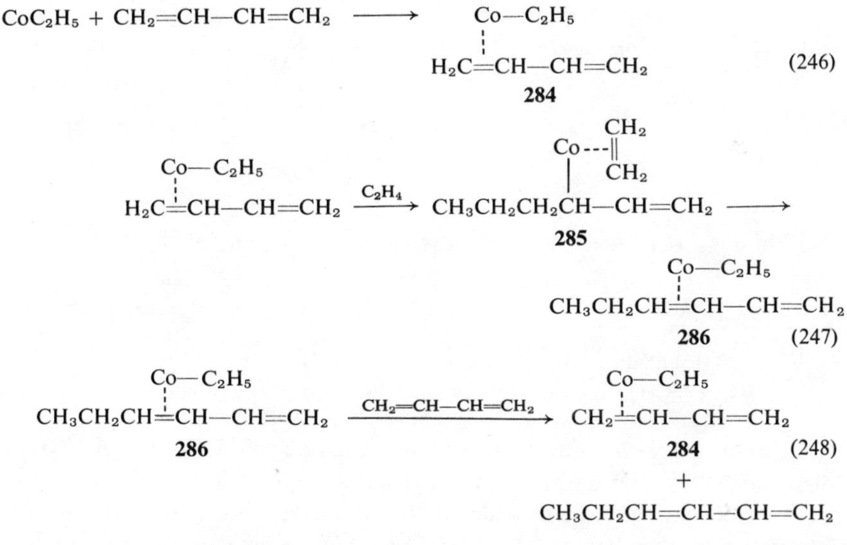

$$CoC_2H_5 + CH_2{=}CH{-}CH{=}CH_2 \longrightarrow$$

$$\begin{array}{c} Co{-}C_2H_5 \\ \vdots \\ H_2C{=}CH{-}CH{=}CH_2 \end{array} \qquad (246)$$
$$\textbf{284}$$

$$\begin{array}{c} Co{-}C_2H_5 \\ \vdots \\ H_2C{=}CH{-}CH{=}CH_2 \end{array} \xrightarrow{C_2H_4} \begin{array}{c} CH_2 \\ Co{-}{-}{-}\| \\ | \quad CH_2 \\ CH_3CH_2CH_2CH{-}CH{=}CH_2 \end{array} \longrightarrow$$
$$\textbf{285}$$

$$\begin{array}{c} Co{-}C_2H_5 \\ \vdots \\ CH_3CH_2CH{=}CH{-}CH{=}CH_2 \end{array}$$
$$\textbf{286} \qquad (247)$$

$$\begin{array}{c} Co{-}C_2H_5 \\ \vdots \\ CH_3CH_2CH{=}CH{-}CH{=}CH_2 \end{array} \xrightarrow{CH_2{=}CH{-}CH{=}CH_2} \begin{array}{c} Co{-}C_2H_5 \\ \vdots \\ CH_2{=}CH{-}CH{=}CH_2 \end{array}$$
$$\textbf{286} \qquad\qquad\qquad \textbf{284} \qquad (248)$$
$$+$$
$$CH_3CH_2CH{=}CH{-}CH{=}CH_2$$

Butadiene reacts with acrylic esters in the presence of zero-valent cobalt catalysts to form 4,6-heptadiene-carboxylic esters [Eq. (249)], probably by a mechanism similar to that suggested above [reactions (243)–(248)] for the formation of 1,3-hexadiene with a catalyst.

$$CH_2{=}CH{-}CH{=}CH_2 + CH_2{=}CH{-}COOR \xrightarrow{Co(0)}$$

$$CH_2{=}CH{-}CH{=}CH{-}CH_2{-}CH_2{-}COOR \qquad (249)$$

Hatta [81] reported the co-oligomerization of butadiene, 1,3-pentadiene, isoprene, and 2,3-dimethyl-1,3-butadiene with ethylene in toluene solution by the use of a mixture of tris(2,4-pentanediono)iron(III) and triethylaluminum as the catalyst (entries 10–13, Table VI). It may be noted from the data in Table VI that, unlike cobalt catalysts, which give mainly 1,3-hexa-diene with butadiene and its derivatives, the iron catalyst preferentially forms stereospecific *cis*-1,4-hexadiene. Formation of the *cis*-1,4-oligomers may be explained with the aid of the reaction sequence (250)–(252). Ethyliron

$$FeL_3 + (C_2H_5)_3Al \rightleftharpoons C_2H_5Fe + C_2H_5AlL_2 + C_2H_4 + HL \tag{250}$$

$$C_2H_5Fe \rightleftharpoons FeH + C_2H_4 \tag{251}$$

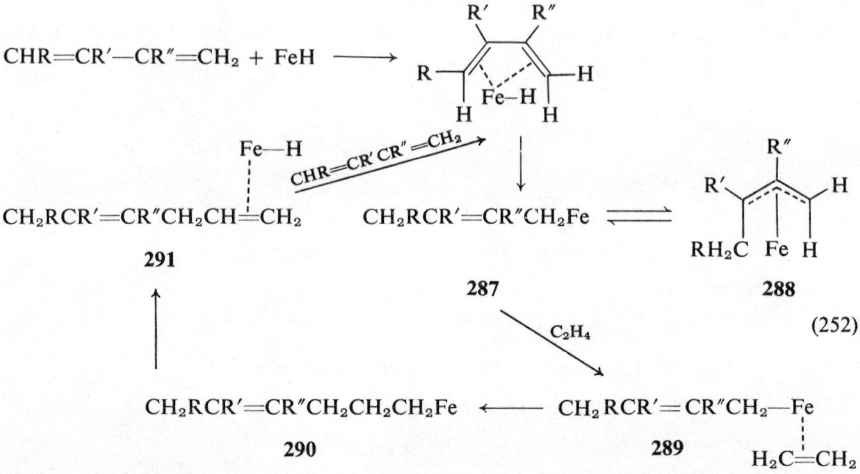

(HL = 2,4-pentanedione;
R=R′=R″=H, butadiene; R=R″=H, R′=CH₃, isoprene;
R′=R″=H, R=CH₃, 1,3-pentadiene; R′=R″=CH₃, R=H, 2,3-dimethyl-1,3-butadiene)

formed by the reduction of tris(2,4-pentanediono)Fe(III) with Al(C₂H₅)₃ dissociates into iron hydride and ethylene in steps (250) and (251), respectively. Addition of iron hydride to the butadiene molecule results in the formation of the dienyl σ-complex **287** which is in equilibrium with the isomeric π-allyl complex **288**. Coordination of ethylene to the σ-dienyl complex yields the intermediate σ-alkenyliron-π-ethylene **289**. Insertion of ethylene then takes place, resulting in the formation of a 1,4-alkenyliron σ complex **290**. Intramolecular hydride transfer in **290** yields a π-bonded intermediate complex **291**, which reacts with another molecule of the diene to form the original σ-alkenyl complex **287** with displacement of the reaction product, 1,4-hexadiene.

It may be noted from the cobalt(0)- or iron(0)-catalyzed co-oligomerization reactions of butadiene described above that formation of a 1,3-diene or a 1,4-diene depends primarily on the olefin insertion steps. When a diene is inserted in a metal–ethyl bond [cobalt(0) catalysis, reaction (247)], 1,3-hexadiene is obtained, whereas insertion of ethylene in a metal–alkenyl bond [iron(0) catalysis, reaction (252)] results in the formation of the 1,4-hexa-dienes.

The reaction of vinylmagnesium chloride with titanium tetrachloride in tetrahydrofuran resulted in the formation of a catalyst that is converted to cyclic and linear dimers of butadiene, along with high molecular weight polymers [290]. Four dimers of butadiene, 4-vinylcyclohexene, 2,6-octadiene, 3-methyl-1,5-heptadiene, and 3-methyl-1,4,6-heptatriene, have been identi-fied in the reaction products by infrared and gas chromatography [290]. The mechanism advanced to account for the formation of these reaction products is outlined in the reaction sequence involving formulas **292–300**.

It has been suggested [290] that tetravinyltitanium **292** is formed in the reaction of the vinyl Grignard reagent with $TiCl_4$. This titanium complex then cleaves homolytically to produce four vinyl radicals with a concerted electron transfer from the vinyl groups to the empty $3d$ orbitals of titanium, reducing it to the zero-valent state. The vinyl radicals then are believed to couple in pairs to form two molecules of 1,3-butadiene by a process of radical hybridization [290,298,299]. The three possible configurations, **294**, **295** and **296**, of the bisbutadiene titanium(0) complex are seen to lead naturally to the four observed oligomers **297**, **298**, **299**, and **300**, with the subsequent separa-tion of titanium metal.

The cyclic dimer 4-vinylcyclohexene, **299**, is preferentially formed from **295** by combination of carbon atoms 1 with 1′ and 2 with 4′. The coupling of carbon atoms 1 to 1′ and 2 to 4′, followed by reduction and rearrangement of hydrogens, gives rise to 2,5-octadiene, **300**, and 3-methyl-1,5-heptadiene, **297**, respectively. 3-Methyl-1,4,6-heptatriene, **298**, is formed by an intramolecular shift of hydrogen atoms with concomitant coupling of carbon atoms 2 and 4′ in **294**.

In a catalytic process, the use of alkyl halide oxidizes the zero-valent titanium obtained by the decomposition of the intermediates **294**, **295**, and **296** to a $Ti(C_nH_{2n-1})X_2$ species which reacts with vinylmagnesium chloride to form **293**. The process thus continues in a catalytic fashion with the forma-tion of 3-methyl-1,5-heptadiene, **297**, and 3-methyl-1,4,6-heptatriene, **298**, along with high molecular weight polymers of butadiene in good yield. The cyclic dimer of butadiene **299**, or the straight-chain isomer, 2,6-octadiene, **300**, are not formed in the reaction.

Olive and Olive [211] have reported the codimerization of butadiene and ethylene to a product consisting of 98% *cis*-1,4-hexadiene with a catalyst consisting of $CoCl_2(diphos)$ and triethylaluminum in 1,2-dichloroethane (entry 29, Table VI). The temperature of the reaction appears to be critical for the production of *cis*-1,4-hexadiene. The high proportion (98%) of *cis*-1,4-hexadiene was formed between 80° and 100°C. Below 80°C, C_8 com-pounds were formed at the expense of the consumption of more ethylene than

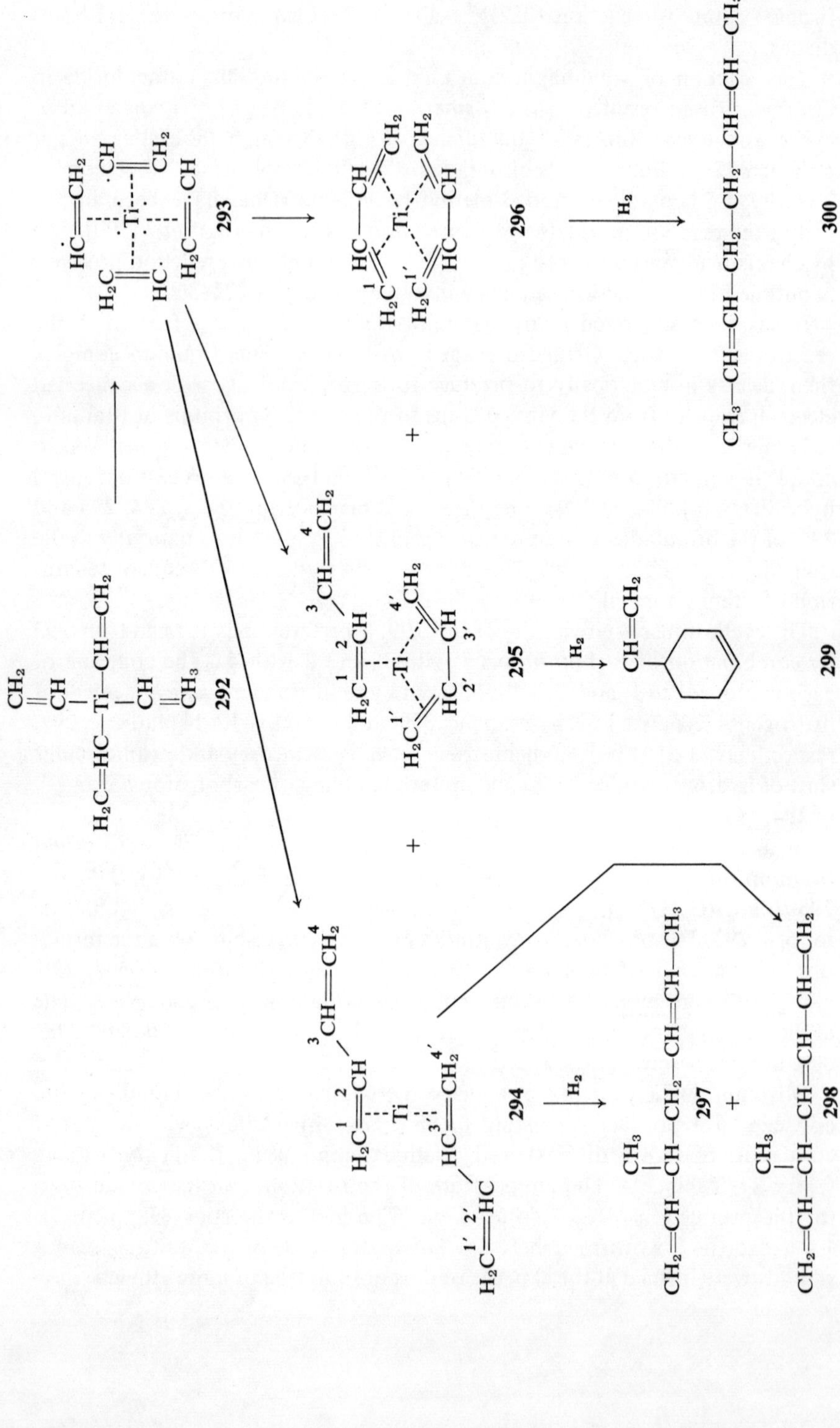

293 $\longrightarrow$ **296** $\xrightarrow{\text{H}_2}$ **300**

$CH_3-CH=CH-CH_2-CH_2-CH_2-CH=CH-CH_3$

+

292

295 $\xrightarrow{\text{H}_2}$ **299**

$CH=CH_2$

+

294 $\xrightarrow{\text{H}_2}$ **297** + **298**

$CH_2=CH-CH_2-CH-CH_2-CH=CH-CH_3$
$\quad\quad\quad\quad\quad | $
$\quad\quad\quad\quad\quad CH_3$

$CH_2=CH-CH_2-CH-CH=CH-CH=CH_2$
$\quad\quad\quad\quad\quad | $
$\quad\quad\quad\quad\quad CH_3$

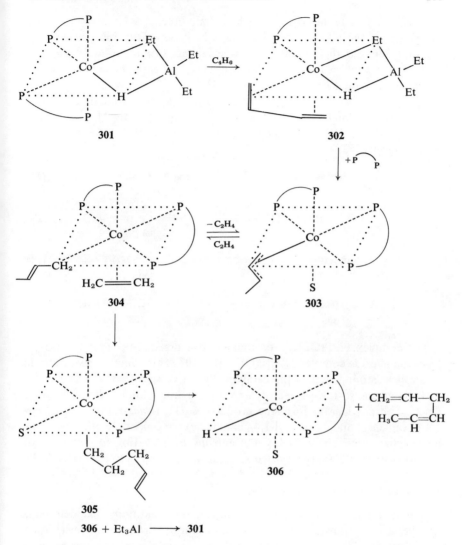

301

302

304

303

305

306

306 + Et₃Al ⟶ 301

butadiene. Above 110°C, 1,4-hexadiene was isomerized to 2,4-hexadiene. The catalytically active species in the reaction has been proposed [211] to be the complex indicated schematically by **301**, with the composition Co(H)-(diphos)₂(Et₃Al).

In the proposed [211] mechanism, complex **301** forms a chelated π complex **302** through displacement of one of the diphosphine ligands by a butadiene molecule. The coordinated butadiene in **302** is converted by hydride transfer to a π-allyl group in **303**. Coordination of ethylene in the σ-butenyl complex

304, followed by its insertion into the cobalt–butenyl σ bond results in the formation of the alkenyl complex **305**, which breaks up into the hydrido complex **306** and the product *cis*-1,4-hexadiene. Combination of **306** with triethylaluminum results in the regeneration of the catalytic complex **301**.

The cationic nickel phosphite complex $HNi[P(OEt)_3]_4{}^+$ catalyzes [212] the coupling of butadiene or ethylene to give a product consisting of 70% *trans*-1,4-hexadiene along with *cis*-1,4-hexadiene and 3-methyl-1,4-pentadiene (entry 30, Table VI). The mechanism proposed [212] for the reaction is illustrated by (253)–(256).

$$H^+ + Ni[P(OEt)_3]_4 \rightleftharpoons HNi[P(OEt_3)]_4{}^+ \tag{253}$$
$$\underset{\textbf{307}}{}$$

$$HNi[P(OEt)_3]_4{}^+ \underset{k_{-1}}{\overset{k_1}{\rightleftharpoons}} HNi[P(OEt)_3]_3{}^+ + P(OEt)_3 \tag{254}$$
$$\underset{\textbf{308}}{}$$

$$HNi[P(OEt)_3]_3{}^+ + C_4H_6 \overset{fast}{\rightleftharpoons} \pi\text{-}C_4H_7Ni[P(OEt)_3]_3{}^+ \tag{255}$$
$$\underset{\textbf{309}}{}$$

$$\pi\text{-}C_4H_7Ni[P(OEt)_3]_3{}^+ + C_2H_4 \longrightarrow C_6H_{10} + HNi[P(OEt)_3]_3{}^+ \tag{256}$$
$$\underset{\textbf{309}}{} \qquad\qquad\qquad \underset{\textbf{308}}{}$$

The complex $Ni[P(OEt_3)]_4$ protonates in a preequilibrium step (253) to form the hydride complex $HNi[P(OEt_3)]_4{}^+$, **307**. This complex dissociates in a rate-determining step to the catalytically active species **308**, which reacts with butadiene to form a π-crotylnickel phosphite complex **309**. The latter reacts with ethylene to form the products C_6H_{10} (*trans*-1,4-hexadiene, *cis*-1,4-hexadiene, and 3-methyl-1,4-pentadiene), regenerating the catalytic intermediate **308**. The enthalpy and entropy of activation for the forward step of reaction (254) are reported as 17 kcal/mole and -10 eu, respectively.

B. CONDITIONS OF OLIGOMERIZATION

It is of interest at this stage to compare the conditions that determine whether a given transition metal catalyst system will produce oligomerization or polymerization. Linear oligomerization or co-oligomerization requires hydrogen transfer from one monomer unit to the other. The example in Eq. (257), suggested by Ochiai [300] for aprotic solvents, illustrated the assistance of hydrogen transfer by coordination of hydride and alkenyl moieties to the metal ion. Although the formal charge on the metal ion may change as the result of this rearrangement, depending on the metal ion and the alkenyl group involved, no change is indicated in the charge of the complex, because of the conventions employed, whereby metal–carbon and metal–hydrogen bonds are written as homopolar linkages.

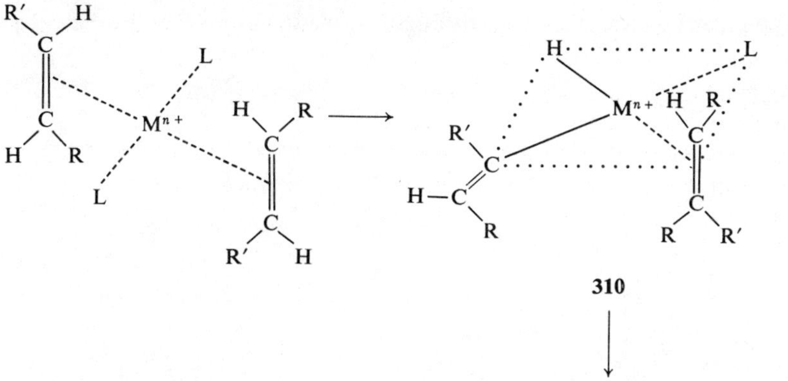

$$(257)$$

310

$$RHC{=}CR'{-}CHR{-}CH_2R' + \text{isomers}$$

In protic solvents a metal–alkyl complex, **311**, is probably first formed by protonation of the olefin complex, followed by an olefin insertion reaction as suggested by Cramer [296]. This mechanism is illustrated by reaction sequence (258).

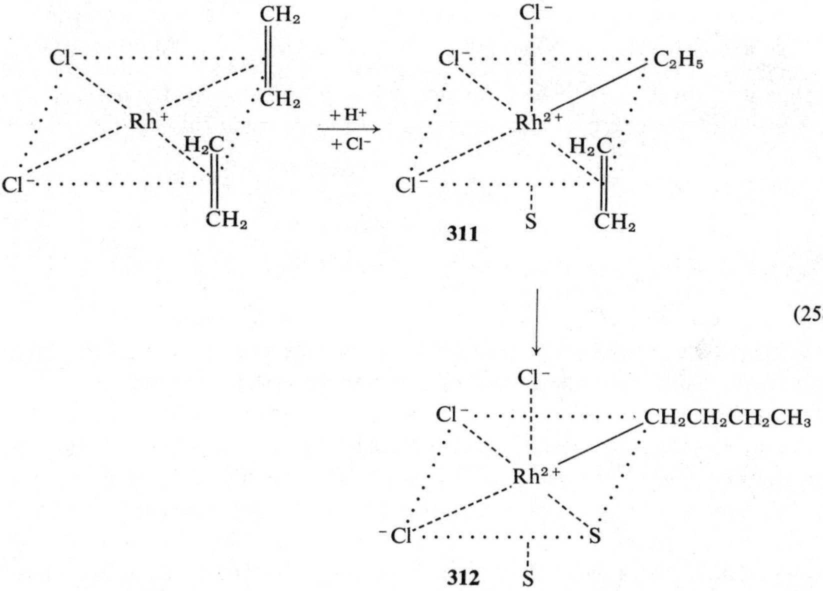

$$(258)$$

311

312

As pointed out by Ochiai [300], a continuous polymerization process is possible only with those catalysts that can maintain the reactivity of the growing chain end. Deactivation of the chain may take place either by a chain transfer to the solvent [reactions (259) and (260)] or by an internal deactivation process in which ends of the chain either combine with each other or with an entering monomer unit, through the mediation of the metal catalyst [reactions (261) and (262)].

1. Chain Transfer to the Solvent

The coordinated monomer or polymer unit may be displaced by a solvent molecule or may be dissociated by reductive elimination.

$$L_5Ti \cdots (CH_2)_nR + S \longrightarrow L_4TiSH + CH_2{=}CH(CH_2)_{n-2}R + L \tag{259}$$

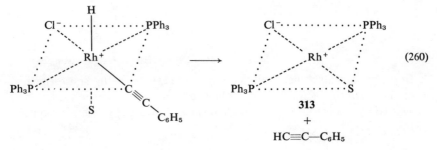

$$313$$
$$+$$
$$HC{\equiv}C{-}C_6H_5$$

2. Chain Transfer to the Monomer Unit

Replacement of the growing chain by monomer terminates the polymerization at the oligomer stage. In processes where recombination of the monomer

$$L_5Ti - (CH_2)_nR + CH_2{=}CH_2 \longrightarrow L_5Ti{-}C_2H_5 + CH_2{=}CH(CH_2)_{n-2}R \tag{261}$$

$$(CH_3CH_2CH{=}CH_2)Cl_2Rh^+S^- + C_2H_4 \longrightarrow C_2H_5Cl_2Rh^+S^- + CH_3CH_2CH{=}CH_2 \tag{262}$$

chain units (deactivation) is faster than, or comparable in rate to, propagation of the chain, oligomerization rather than polymerization occurs.

In the examples given [Eq. (258)–(262)] it is seen that the catalytically active species for oligomerization processes are metal complexes in which the metal ions are in their lowest valence state. Thus cobalt and iron complexes in their zero-valent or (-1) oxidation states catalyze linear dimerization of olefins and butadiene, because they can form kinetically labile metal hydride intermediates that are capable of transferring hydrogen to the coordinated monomer units. In their $+1$, or higher-valent, oxidation states, however, iron and cobalt catalyze continuous polymerization of olefins or butadiene. Zero-valent nickel can take part in oligomerization of butadiene, but in the $+2$ oxidation state (e.g., $[Ni(\pi\text{-allyl})_2Br]_2$) nickel catalyzes polymerization of butadiene [301]. Rhodium(I), palladium(II), ruthenium(II), and platinum(II) complexes oligomerize butadiene in nonaqueous solvents. Rhodium(III) in the presence of a coordinated halide ion catalyzes dimerization of olefins in aqueous and alcoholic solvents. Chloride-free rhodium(III) salts, however, are very effective polymerization catalysts, especially in dilute aqueous solutions. Cramer [296] has suggested that the presence of chloride ions in the coordination sphere of rhodium(III) stops polymerization by a chain-transfer process to the solvent (deactivation of the chain).

C. LINEAR OLIGOMERIZATION OF ALKYNES

Tris(triphenylphosphine)chlororhodium(I) in benzene, methylene chloride, or chloroform, catalyzes the conversion of α-hydroxyacetylenes predominantly to dimers [213], as indicated by reaction (263). 3-Methyl-1-butyne-3-ol

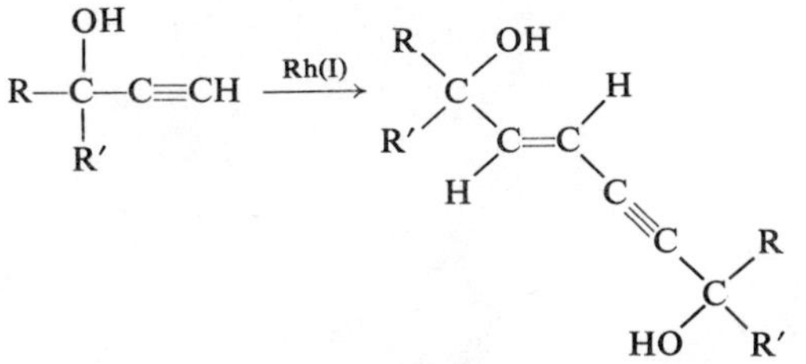

$$
R-\underset{\underset{R'}{|}}{\overset{\overset{OH}{|}}{C}}-C\equiv CH \xrightarrow{Rh(I)} \quad \text{(263)}
$$

314

(R = aliphatic group; R′ = aliphatic group or hydrogen)

gave the dimer 2,7-dimethyl-oct-3-en-5-yne-2,7-diol, which was characterized by its infrared and NMR spectra. The catalysts $RuCl_2(PPh_3)_2$ and $RhCl(CO)(PPh_3)_2$ are not effective in the dimerization of the alkyne. With phenylacetylene, the catalyst **313** induces extensive polymerization quantitatively at about 50°C in less than 10 hours.

Since the acetylated derivative of 3-methyl-1-butyne-3-ol was not dimerized with the rhodium complex, it was concluded that the α-hydroxy group of the alkyne is essential for dimerization. In the mechanism illustrated by chemical transformations of **315** to **319** and **320**, the combination of π-bonded and σ-bonded acetylene derivatives via a three center transition state **317** was postulated to explain dimer formation [213].

It seems probable that the α-hydroxyl group of the coordinated alkyne enhances the stability of the intermediates **316**, **317**, and **318** by partial hydrogen bond formation with the coordinated chloride, thus facilitating the transfer of the acetylide residue to the coordinated alkyne in the transition state. The hydroxyl group may also discourage further growth of the chain by facilitating chain transfer to the solvent as illustrated in the last step (**318 → 319,320**). Phenylacetylene itself, without an α-hydroxyl group, gives only high-polymeric products [278].

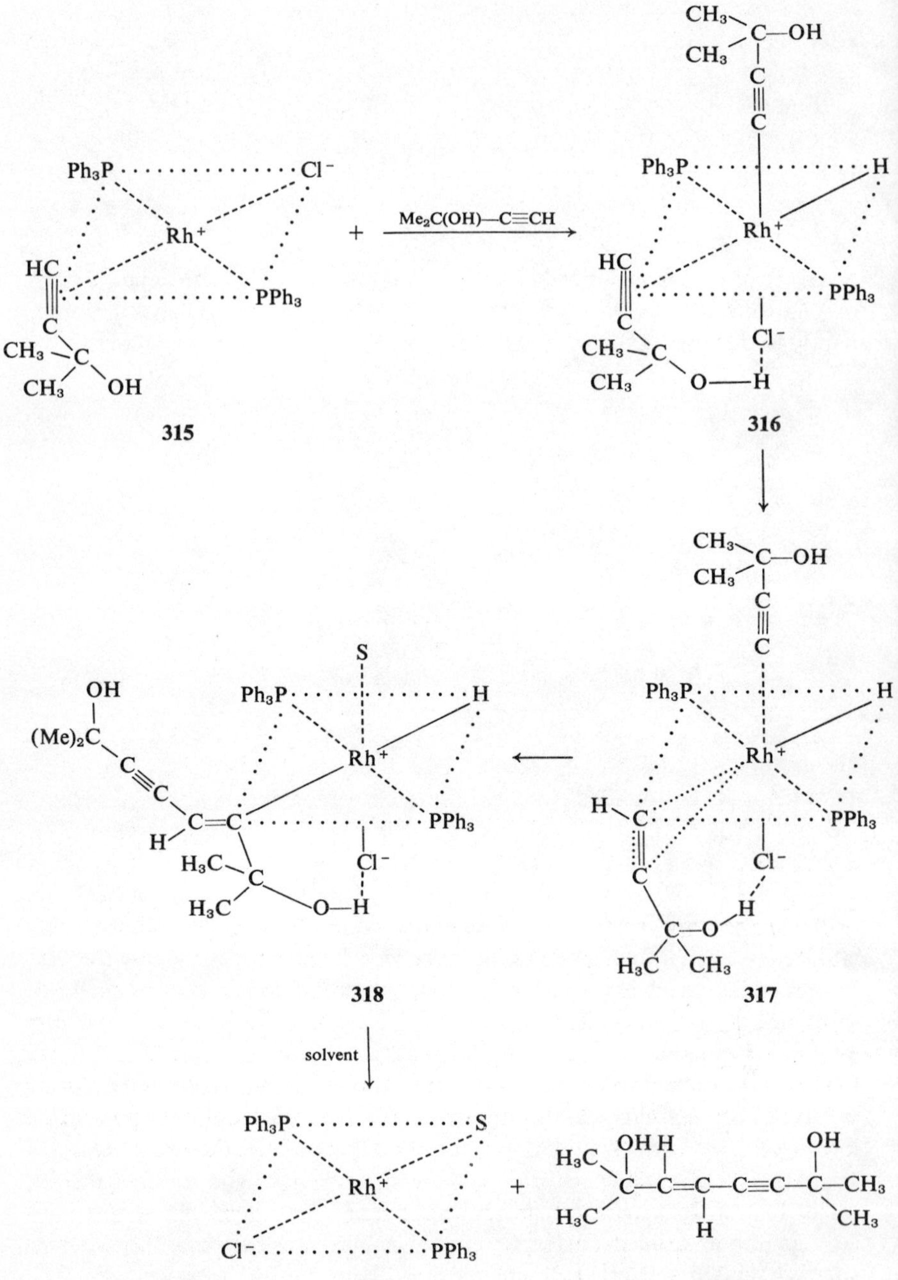

315 **316** **317** **318** **319** **320**

D. CYCLIC OLIGOMERIZATION AND CO-OLIGOMERIZATION OF DIENES

The majority of the cyclic oligomerization reactions of butadiene that have been reported have been carried out with catalyst systems containing nickel(0), iron(0), and cobalt(0) complexes. Dimerization of butadiene to 1,5-cyclooctadiene, **322**, has been accomplished in 95% yield by the use of diethylbis(dipyridyl)iron [277], **321**, as indicated by reaction (264). With butadiene-1,1,4,4-d_4, cyclooctadiene-3,3,4,4,7,7,8,8-d_8, **324**, and 4-vinyl-d_2-cyclohexene-3,3,5,5,6,6,d_6, **325**, were obtained [reaction (265)].

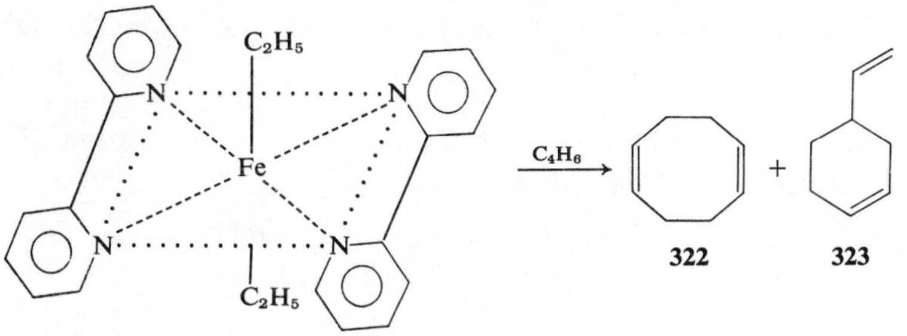

$$ \text{321} \xrightarrow{C_4H_6} \text{322} + \text{323} \qquad (264) $$

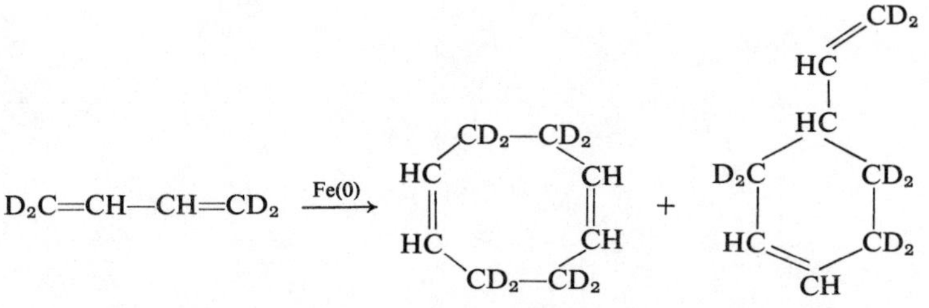

$$ D_2C{=}CH{-}CH{=}CD_2 \xrightarrow{Fe(0)} \text{324} + \text{325} \qquad (265) $$

Zero-valent nickel catalysts that are very effective in the dimerization and trimerization of butadiene are bis(1,5-cyclooctadiene)nickel(0), **326**, bis(cyclooctatetraenenickel(0), **327**, and 1,5,9-cyclododecatrienenickel(0) **328**, obtained readily by the reduction of bis(2,4-pentadiono)Ni(II) with aluminum alkyl in the presence of the corresponding hydrocarbons at low temperatures [302,303]. Spectral evidence on complex **326** indicates the coordination of all the double bonds to the nickel atom. The geometric conformation of nickel in complex **326** is not yet certain. Complex **327** is black, sparingly soluble (in organic solvents) and is probably polymeric. Complex **328**, the *trans,trans, trans*-cyclododecatriene-Ni(0) complex, crystallizes as deep red needles and may be sublimed in vacuum. It is monomeric in the vapor phase and in solution, and is extremely air-sensitive.

Butadiene oligomerizes in the presence of **326** and **328** at 20°C to form a trimer, cyclododecatriene, possessing mainly the *trans,trans,cis* configuration. Small amounts of the *trans,cis,cis* isomer of the cyclic triene have also been obtained. When the polymerization of butadiene is conducted at −40°C with **326** or **328** as catalyst, an open-chain product, **329**, with composition $C_{12}H_{18}$-Ni was isolated from the reaction mixture. This complex contains a coordinated *trans* double bond and two terminal coordinated π-allyl groups. On hydrogenation of **329**, a quantitative yield of *n*-dodecane is obtained. Ring closure of **329** to π-bonded cyclododecatriene-Ni(0) complexes **328** and **330** takes place either on heating to 20°C or by treatment with ligands such as carbon monoxide, phosphines, and butadiene. When reaction of **329** with carbon monoxide is carried out at −60°C, a molecule of carbon monoxide is inserted into the carbon skeleton to give a vinylcycloundecadienone, **331**, with displacement of the nickel as nickel carbonyl.

The structure and reaction types assigned to **329** are supported by reactions

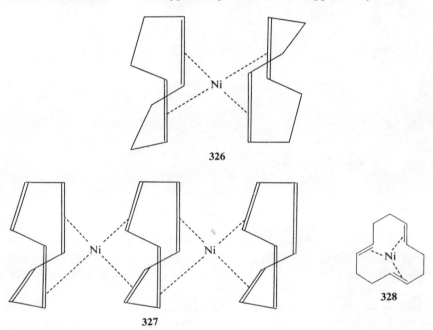

326

327

328

of a related complex, bis(π-allyl)nickel(0), **332**, synthesized by the reaction of allylmagnesium bromide with anhydrous nickel(II) bromide in ether. The compound crystallizes below 1°C to long yellow needles soluble in ether. When dissolved in ether, complex **332** is stable towards water, in accord with its π-complex structure. The structure has been confirmed by X-ray analysis to consist of two parallel *trans*-π-allyl planes at right angles to the nickel axis and equidistant from the metal. Reduction of **332** with hydrogen yields

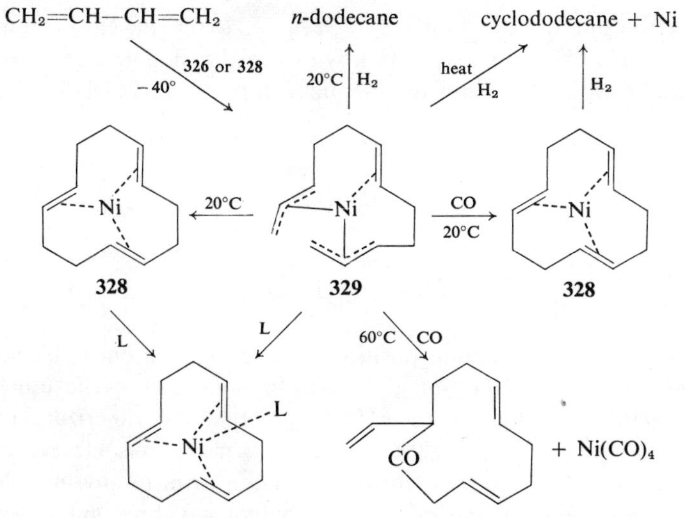

$$CH_2{=}CH{-}CH{=}CH_2 \qquad n\text{-dodecane} \qquad cyclododecane + Ni$$

(L = PR$_3$, C$_4$H$_6$)

propane and nickel. Reaction with carbon monoxide or a phosphine causes coupling of the two allyl ligand radicals to form diallyl, **333**. Reaction of **332** with butadiene resulted in the polymerization of butadiene to cyclododeca-trienenickel(0), **328**, in addition to the formation of **333**. The analogous reactions and properties of **329** and **332** thus support the open-chain structure assigned to **329**.

Complex **329** thus seems to be an appropriate model for intermediates in the cyclization reaction that takes place on zero-valent nickel [303]. The rearrangement of bonds in the conversion of **329** to **328** takes place by an electron shift with rehybridization of terminal carbon atoms from sp^2 to sp^3. According to Schrauzer [304,305] the driving force for the electronic re-arrangement is the tendency of the metal complex system in **329** to achieve higher orbital stabilization in the cyclododecatrienenickel(0) complex, **328**.

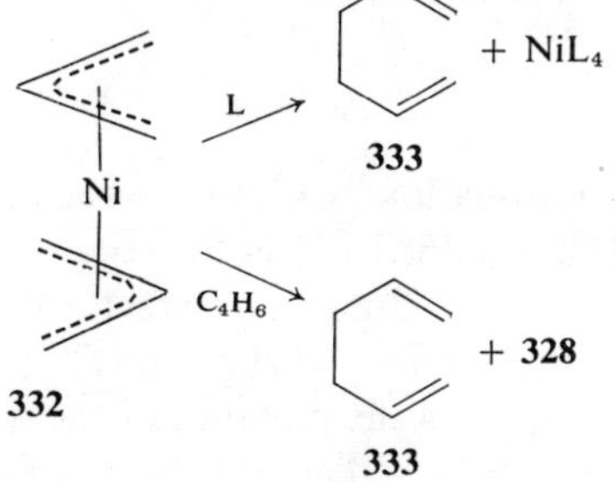

(266)

(267)

(L = CO or PR$_3$)

The electron shift indicated for **329** → **329a** → **328** for ring closure apparently is facilitated by heat, or by reaction with coordinating groups that provide additional coordination sites for the nickel atom [Eq. (268)].

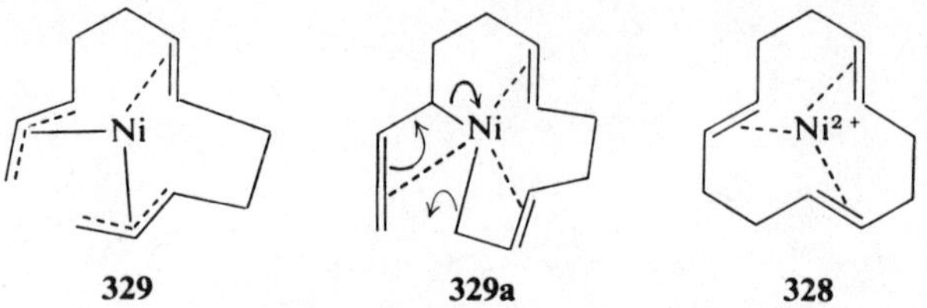

$$ \tag{268} $$

329   **329a**   **328**

If one of the coordination positions of the nickel atom is blocked by a molecule of tertiary phosphine, butadiene undergoes cyclic dimerization to [Eq. (269)] 1,5-cyclooctadiene, **335**, rather than the trimerization reaction described above [303]. A catalyst for cyclic dimerization is tetrakis(triphenyl-phosphine)nickel(0) obtained from the reaction of bis(π-allyl)nickel(0) with triphenylphosphine. In the presence of butadiene, three molecules of the phosphine are displaced, while one is retained. At 80°C a yield of product containing over 95% cyclooctadiene has been obtained with a relatively high reaction rate.

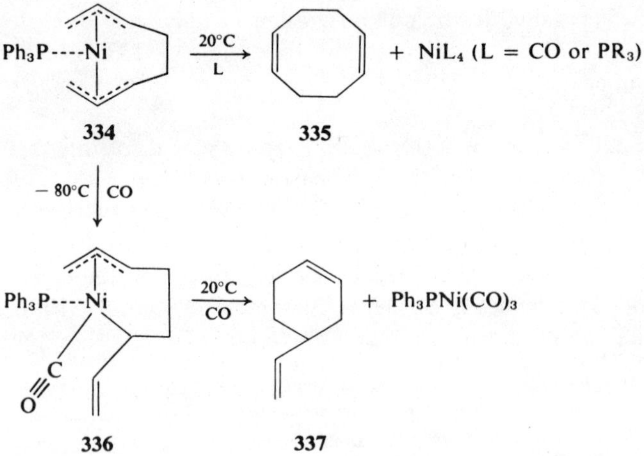

$$ Ph_3P \cdots Ni \quad \xrightarrow[L]{20°C} \quad \bigcirc \quad + \quad NiL_4 \text{ (L = CO or } PR_3) \tag{269} $$

334   **335**

$$ -80°C \downarrow CO $$

$$ Ph_3P \cdots Ni \quad \xrightarrow[CO]{20°C} \quad + \quad Ph_3PNi(CO)_3 \tag{270} $$

336   **337**

Since one of the coordination positions on the nickel atom is blocked, the original complex, **334**, obtained by the reaction of $Ni[(C_6H_5)_3P]_4$ with two molecules of butadiene is a "dimer" in which two allyl groups are bonded together and coordinated to the nickel atom. The intermediate **336** has been isolated at low temperature in the presence of excess carbon monoxide. Ring closure of **334** and **336** is achieved at 20°C in the presence of donor molecules such as carbon monoxide or trialkylphosphine to form 1,5-cyclooctadiene and 4-vinylcyclohexene, **335** and **337**, respectively. Vinylcyclohexene, **337**, is formed in yields of more than 81% if **334** is treated with excess of carbon monoxide at a low temperature. Coordination with carbon monoxide breaks up one of the π-allyl groups, in **334**. Electron shifts coupled with carbon–carbon bond formation then take place with subsequent dissociation of vinylcyclohexene.

Cyclo-co-oligomerization of butadiene and ethylene with nickel(0) catalysts at 20°C and 20–30 atm gives *cis,trans*-1,5-cyclododecadiene, **338**, in 80% yield [reaction (271)]. Cyclo-co-oligomerization of butadiene and 2-butyne resulted in the formation of 1,2-dimethyl-1,4,8-cyclodecatriene **340** [reaction (272)]. On heating, **338** and **340** undergo the Cope rearrangement to form 1,2-divinylcyclohexane, **339**, and dimethyldivinylcyclohexene, **341**, respectively.

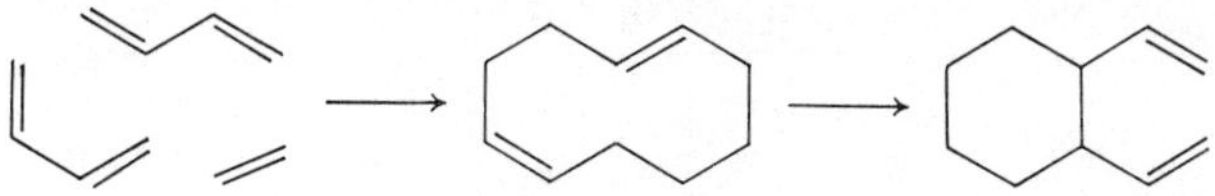

(271)

338   **339**

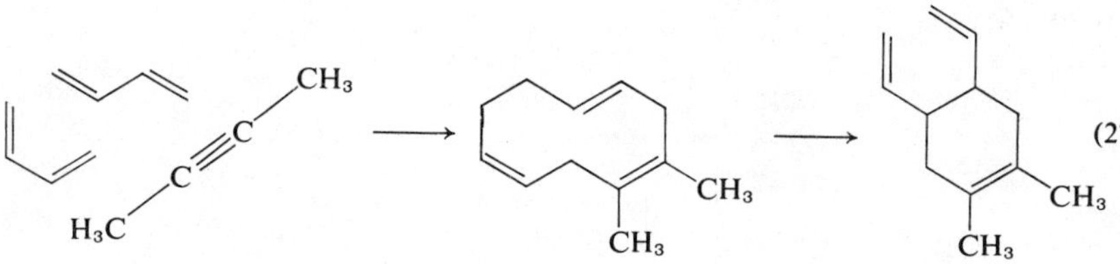

(272)

340   **341**

VII. Cyclic Oligomerization of Alkynes

Catalytic cyclic trimerization and tetramerization of acetylene provide synthetic routes to benzene and cyclooctatetraene [306]. Cyclic tetramerization of acetylene was first carried out heterogeneously by Reppe [90,264] with a metallic nickel catalyst, thus providing a relatively inexpensive synthesis of cyclooctatetraene in commercial quantities. The first homogeneous nickel(0) catalyst developed for the synthesis of cyclooctatetraene was bis(acrylonitrile)nickel(0) [307,308] with which a 70% yield of the hydrocarbon was obtained. Small quantities of benzene, styrene, vinylcyclooctatetraene, and 1-phenyl-1,3-butadiene were also formed.

Effective catalysts for the cyclic tetramerization of alkynes have been obtained from nickel(II) complexes of 2,4-pentanedione, salicylaldehyde, *N*-alkylsalicylaldimines, and cyanide [304]. The nickel(II) complexes that show catalytic activity in the conversion of acetylene to cyclooctatetraene are generally octahedral and paramagnetic. Planar nickel(II) complexes that are converted to the high-spin form through the action of polar solvents have also been observed to be active catalysts. The reaction proceeds more rapidly in tetrahydrofuran and dioxane than in benzene. Strongly polar solvents such as water, pyridine, and benzonitrile are not suitable for the cyclic oligomerization of alkynes, since such solvents are not readily displaced by alkynes from the coordination sphere of the metal ions. Suitable substrates are acetylene and monosubstituted acetylenes. The mechanism in Eq. (273) has been proposed for the cyclic oligomerization of alkynes [304].

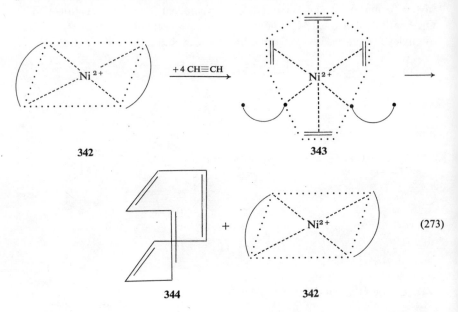

The chelate nickel(II) complex **342** is visualized as reacting with four molecules of the alkyne with cleavage of the chelate ring to form an octahedral complex, **343**, in which the alkyne molecules assume a configuration around the nickel(II) ion favorable to the formation of cyclooctatetraene. In cases where bis(N-alkylsalicylaldiminato)nickel(II) catalysts are used, the yields of cyclooctatetraene vary with the ligand-field strength. The maximum yield of the tetramer was obtained with N-methyl Schiff base chelates, which form the most highly paramagnetic complexes. The paramagnetism of octahedrally coordinated nickel(II) thus seems to strongly influence, or at least parallels, the coupling of acetylene or alkynes in the transition state. In the intermediate octahedral complex **343**, π-bond formation with the nickel(II) ion may assist the hybridization of sp carbon atoms to sp^2, and imparts to each alkyne molecule partial biradical characteristics. The formation of cyclooctatetraene does not seem to involve any intermediate, although the formation of all four new C—C bonds may not be a concerted process.

Support for the mechanistic picture developed in Eq. (273), in which acetylene is activated through formation of a complex of the type indicated by **343**, is provided when one of the active coordination sites on nickel(II) is occupied by an additional unidentate ligand such as triphenylphosphine formula **345**. In such cases, formation of cyclooctatetraene does not occur, and instead a quantitative trimerization of alkyne or acetylene to benzene derivatives is the only possible simple oligomerization reaction. Occupation of one coordination position of nickel(II) by triphenylphosphine thus favors cyclic trimerization, according to the reaction scheme (274).

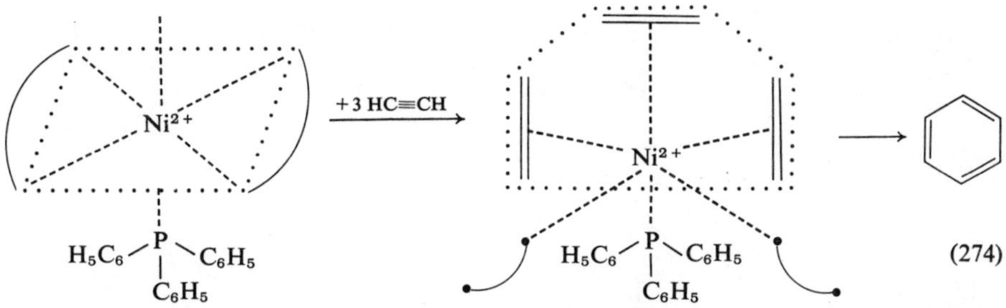

$$\text{(274)}$$

345   **346**

Bidentate ligands such as α,α-bipyridyl and 1,10-phenanthroline poison the Ni(II) catalyst by occupying a larger number of coordination positions, thus making impossible the steric organization of acetylene moieties necessary for the formation of cyclic trimers and tetramers, as indicated in reaction sequences (273) and (274). The formation of small quantities of styrene, phenylbutadiene, and vinylcyclooctatetraene in the normal trimerization and tetramerization processes may be explained on the basis of a mixed cyclo-co-oligomerization reaction of alkynes with vinylacetylene and higher linear oligomers of acetylene.

An alternate mechanism for the cyclic trimerization of alkynes, originally proposed by Meriwether et al. [276], is the participation of metallocyclobutadiene intermediates in the reaction. Such a possibility has been given considerable support by the recent work of Collman et al. [309,310] on the iridium(I)-catalyzed cyclotrimerization of alkynes to benzene derivatives. The iridium(I)–dinitrogen complex **347** reacts with a molecule of dimethylacetylene dicarboxylate to form an iridocyclopropene derivative **348** with the loss of nitrogen. This complex reacts with another molecule of the alkyne, to yield a metallocyclobutadiene complex **349** in which octahedral iridium is coordinatively unsaturated and binds either a molecule of carbon monoxide

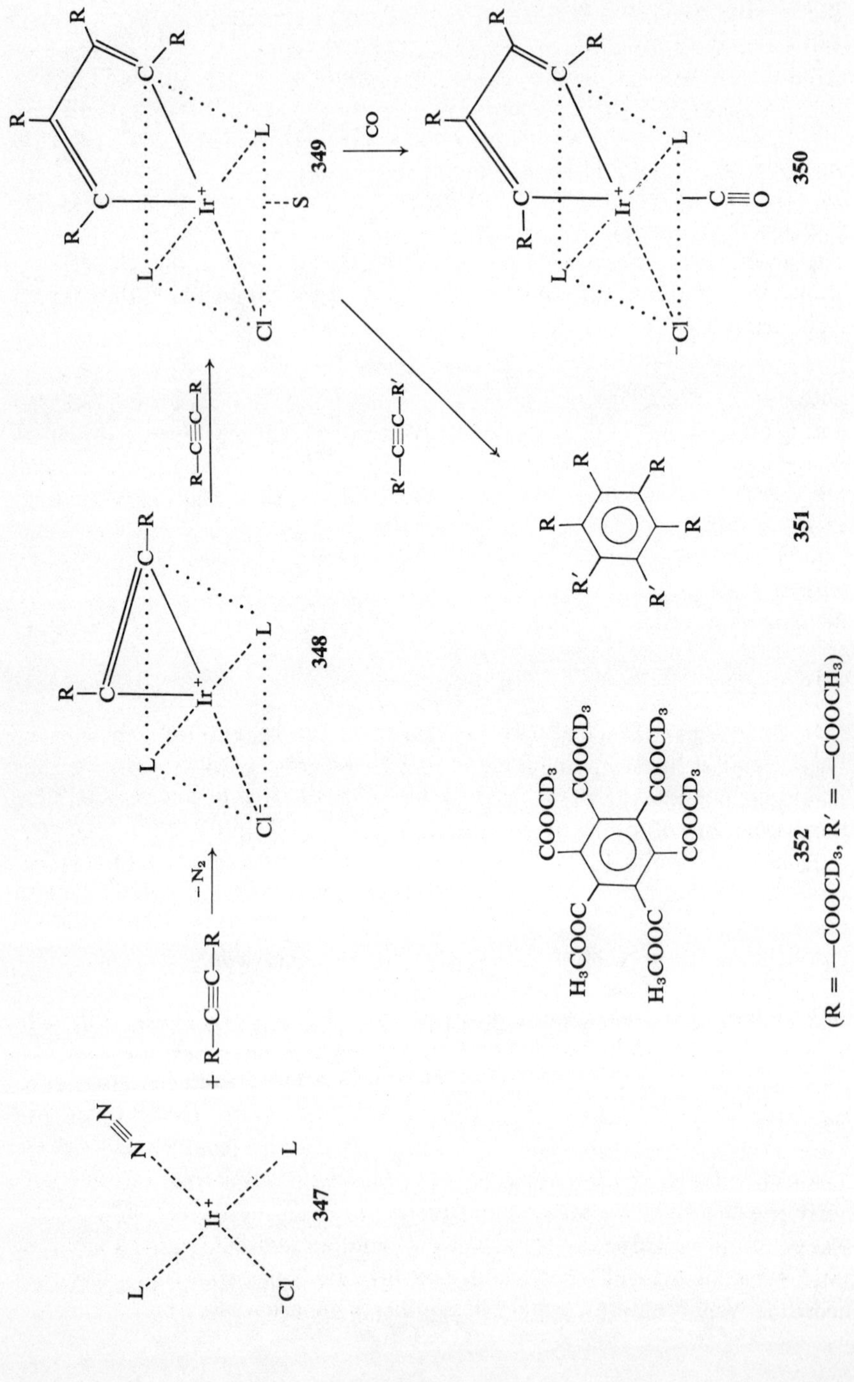

349

$\xrightarrow{\text{CO}}$

350

R—C≡C—R

R'—C≡C—R'

348

351

+ R—C≡C—R $\xrightarrow{-N_2}$

352

347

(R = —COOCD₃, R' = —COOCH₃)

to form the carbonyl complex **350**, or catalyzes the trimerization of the alkyne above 100°C to the substituted benzene **351**. The name "iridocycle" has been tentatively suggested for complexes **349** and **350** by Collman *et al.* [309]. A vacant coordination site in **349** is apparently essential for the catalytic trimerization of alkyne. Participation of **349** in the trimerization of the alkyne is further supported by the interaction of the perdeuterioiridocycle with a mole of dimethylacetylene dicarboxylate to form dodecadeuterohexacarbomethoxybenzene, **352**, and a protioiridocycle.

Recently, Reinheimer *et al.* [311] described the Pd(II)-catalyzed trimerization of 2-butyne to hexamethylbenzene through a novel intermediate species having the composition $[(Me_2C_2)_3PdCl_2]_2$. The structure now suggested for this complex, illustrated by **353**, is considered to have organic ligands designated as 2-chloro-3,4,5,6-tetramethyl-2-*trans*-4-*cis*-6-*cis*-octatriene. This ligand is pictured as being σ-bonded to palladium(II) at the 7-position and π-bonded to the palladium at the 2-3-position. This intermediate has been prepared by the interaction of dichlorobis(benzonitrile)palladium(II) and an excess of 2-butyne at 10°C. The composition of the complex has been confirmed by elemental analysis, and the structure indicated is in conformity with its infrared and NMR spectra. This complex is converted to hexamethylbenzene and palladium dichloride at 34°C in chloroform solution.

$$6\ H_3C\!-\!C\!\equiv\!C\!-\!CH_3 + 2\ (C_6H_5CN)_2PdCl_2 \longrightarrow$$

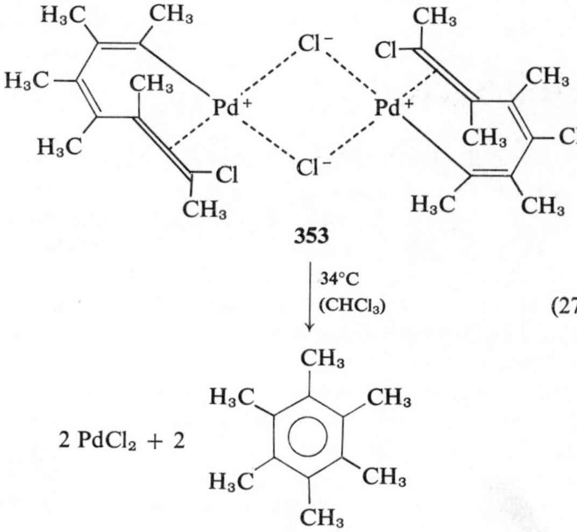

353

(275)

Further studies of the mechanistic details of this reaction would provide interesting information about the nature of insertion reactions of coordinated acetylene in the Pd—C bond to give oligomers and high polymers.

It has recently been suggested by Reinheimer *et al.* [312] and Maitlis [200] that complex **353** is formed from 2-butyne and palladium(II) chloride by the mechanism shown in Eq. (276). The reaction proceeds by the coordination of

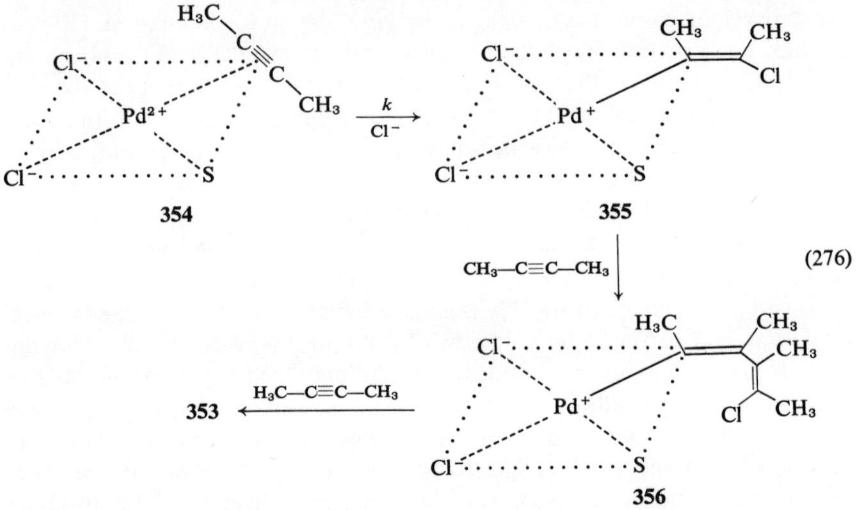

$$\text{CH}_3\text{—C}\equiv\text{C—CH}_3$$

(276)

one molecule of 2-butyne to $PdCl_2$ to form **354** which rearranges to **355** in a slow step. Fast *cis* insertion of 2-butyne in **355** gives **356**, which gives the final product **353** by another *cis* insertion of 2-butyne.

Bibliography

1. M. J. S. Dewar, *Bull. Soc. Chim. Fr.* [5] **18C**, 79 (1951).
2. J. Chatt and L. A. Duncanson, *J. Chem. Soc., London* p. 2939 (1953).
3. J. A. Wunderlich and D. P. Mellor, *Acta Crystallogr.* **7**, 130 (1954).
4. J. A. Wunderlich and D. P. Mellor, *Acta Crystallogr.* **8**, 57 (1955).
5. A. I. Gusev and Yu. T. Struchkov, *J. Struct. Chem. (USSR)* **11**, 340 (1970).
6. R. Denning, F. R. Hartley, and L. M. Venanzi, *J. Chem. Soc., A* p. 324 (1967).
7. C. E. Halloway, G. Hulley, B. F. G. Johnson, and J. Lewis, *J. Chem. Soc., A* p. 53 (1969).
8. F. R. Hartley, *Chem. Rev.* **59**, 799 (1969).
9. H. B. Gray and C. J. Ballhausen, *J. Amer. Chem. Soc.* **85**, 260 (1963).
10. G. R. Davies, W. Hertson, R. H. B. Mais, and P. G. Owaston, *Chem. Commun.* p. 423 (1967).
11. J. H. Nelson, N. S. Wheelock, L. C. Gusachs, and H. B. Jonassen, *J. Amer. Chem. Soc.* **91**, 7005 (1969).
12. J. H. Nelson and H. B. Jonassen, *Coord. Chem. Rev.* **6**, 2763 (1971).
13. J. H. Nelson, K. S. Wheelock, L. C. Cusachs, and H. B. Jonassen, *Inorg. Chem.* **11**, 422 (1972).
14. M. Orchin, *Advan. Catal.* **16**, 1 (1966).

15. I. Wender, H. W. Sternberg, and M. Orchin, *J. Amer. Chem. Soc.* **75**, 3041 (1953).
16. I. Wender, S. Metlin, S. Ergun, H. W. Sternberg, and H. Greenfield, *J. Amer. Chem. Soc.* **78**, 5401 (1956).
17. R. F. Heck and D. S. Breslow, *J. Amer. Chem. Soc.* **83**, 4023 (1961).
18. G. L. Karapinka and M. Orchin, *J. Org. Chem.* **26**, 4187 (1961).
19. M. Johnson, *J. Chem. Soc., London* p. 4859 (1963).
20. H. W. Sternberg and I. Wender, *Chem. Soc., Spec. Publ.* **13**, 53 (1959).
21. P. Taylor and M. Orchin, *J. Amer. Chem. Soc.* **93**, 6504 (1971).
22. H. Rehlen, A. Gruhl, G. Hessling, and D. Pfrengle, *Justus Leibigs Ann. Chem.* **482**, 11 (1930).
23. B. F. Hallam and P. F. Pauson, *J. Chem. Soc., London* **88**, 2272 (1966).
24. E. O. Fischer and H. Werner, *Angew. Chem., Int. Ed. Engl.* **2**, 80 (1963).
25. M. L. G. Green and P. L. K. Nagy, *Advan. Organometal. Chem.* **2**, 325 (1964).
26. K. G. Guy and B. L. Shaw, *Advan. Inorg. Chem. Radiochem.* **4**, 115 (1962).
27. T. A. Manuel, *J. Org. Chem.* **27**, 3941 (1962).
28. W. Heiber, W. Beck, and G. Braun, *Angew. Chem.* **72**, 795 (1960).
29. T. A. Manuel, *Trans. N.Y. Acad. Sci.* [2] **26**, 442 (1964).
30. H. W. Sternberg, R. Markby, and I. Wender, *J. Amer. Chem. Soc.* **78**, 5704 (1956).
31. H. W. Sternberg, R. Markby, and I. Wender, *J. Amer. Chem. Soc.* **79**, 6116 (1957).
32. H. D. Murdoch and E. Weiss, *Helv. Chim. Acta* **48**, 1927 (1962).
33. R. Pettit, G. Emerson, and J. Maher, *J. Chem. Educ.* **40**, 175 (1963).
34. G. F. Emerson and R. Pettit, *J. Amer. Chem. Soc.* **84**, 4591 (1962).
35. L. Roos and M. Orchin, *J. Amer. Chem. Soc.* **87**, 5502 (1965).
36. R. B. King, T. A. Manuel, and F. G. A. Stone, *J. Inorg. Nucl. Chem.* **16**, 233 (1961).
37. T. A. Manuel and F. G. A. Stone, *J. Amer. Chem. Soc.* **82**, 366 (1960).
38. J. E. Arnet and R. Pettit, *J. Amer. Chem. Soc.* **83**, 2954 (1961).
39. R. Pettit and G. Emerson, *Advan. Organometal. Chem.* **1**, 1 (1964).
40. R. Pettit, *Ann. N. Y. Acad. Sci.* **125**, 89 (1965).
41. G. F. Emerson, J. E. Mahler, and R. Pettit, *Chem. Ind. (London)* p. 836 (1964).
42. I. Ogata and A. Misono, *J. Chem. Soc. Jap., Pure Chem. Sect.* **85**, 753 (1964).
43. R. E. Rinehart and J. S. Lasky, *J. Amer. Chem. Soc.* **86**, 2516 (1964).
44. J. E. Mahler and R. Pettit, *J. Amer. Chem. Soc.* **85**, 3955 (1963).
45. J. E. Mahler, D. G. Gibson, and R. Pettit, *J. Amer. Chem. Soc.* **85**, 3959 (1963).
46. F. J. Impastato and K. G. Ihrman, *J. Amer. Chem. Soc.* **83**, 3726 (1961).
47. J. J. Rooney and G. Webb, *J. Catal.* **4**, 1 (1956).
48. J. F. Harrod and A. J. Chalk, *J. Amer. Chem. Soc.* **86**, 1776 (1964).
49. G. Bond and M. Hellier, *Chem. Ind. (London)* p. 35 (1965).
50. G. Bond and M. Hellier, *J. Catal.* **4**, 1 (1965).
51. M. B. Sparke and L. Turner, Belgian Patent 612,300 (1962).
52. M. B. Sparke and L. Turner, *Int. Union Pure Appl. Chem., Abstr. Div. A* Paper A, B 4–30, p. 175 (1963).
53. M. B. Sparke, L. Turner, and A. J. M. Wenham, *J. Catal.* **4**, 332 (1965).
54. E. W. Stern, *Proc. Chem. Soc., London* p. 111 (1963).
55. E. W. Stern, *Catal. Rev.* **1**, 73 (1967).
56. R. Cramer and R. V. Lindsey, *J. Amer. Chem. Soc.* **88**, 3534 (1966).
57. N. R. Davies, *Nature (London)* **201**, 490 (1964).
58. J. F. Harrod and A. J. Chalk, *J. Amer. Chem. Soc.* **88**, 3491 (1966).
59. R. Cramer, *J. Amer. Chem. Soc.* **88**, 2272 (1966).
60. J. Milogram and W. N. Urey, *Proc. Int. Conf. Coord. Chem., 7th, 1962 Abstracts*, p. 264 (1962).

61. H. A. Tayim and J. C. Bailar, Jr., *J. Amer. Chem. Soc.* **89**, 3420 (1967).
62. J. C. Trebellas, J. R. Olechowski, H. B. Honassen, and D. W. Moore, *J. Organometal. Chem.* **9**, 153 (1967).
63. J. W. Rowe, *Proc. Chem. Soc., London* p. 66 (1962).
64. M. L. H. Green and P. L. K. Nagy, *J. Amer. Chem. Soc.* **84**, 1310 (1962).
65. N. R. Davies, *Aust. J. Chem.* **17**, 212 (1964).
66. J. F. Harrod and A. J. Chalk, *Nature (London)* **205**, 280 (1965).
67. B. Cruikshank and N. R. Davies, *Aust. J. Chem.* **19**, 815 (1966).
68. J. Kovacs, G. Speier, and L. Marko, *Inorg. Chim. Acta* **4**, 412 (1970).
69. C. T. Kisker and D. J. Crandale, *Tetrahedron Lett.* No. 19, p. 701 (1963).
70. O. Roelen, German Patent R103,362 (1938).
71. O. Roelen, U.S. Patent, 2,327,066 (1943).
72. O. Roelen, *Angew. Chem.* **60**, 213 (1948).
73. O. Roelen, *Angew. Chem.* **63**, 482 (1951).
74. I. Wender, H. W. Sternberg, R. A. Freidel, J. Metlin, and R. Markby, *U.S., Bur. Mines, Bull.* **600** (1962).
75. H. Wakamatsu, *J. Chem. Soc. Jap., Pure Chem. Sect.* **85**, 227 (1964).
76. J. B. Zachry, *Ann. N.Y. Acad. Sci.* **125**, 154 (1965).
77. F. H. Jardine, J. A. Osborn, G. Wilkinson, and J. F. Young, *Chem. Ind. (London)* p. 560 (1965).
78. J. A. Osborn, F. H. Jardine, J. F. Young, and G. Wilkinson, *J. Chem. Soc., A* p. 1711 (1965).
79. G. Yagupsky, C. K. Brown, and G. Wilkinson, *J. Chem. Soc., A* p. 1392 (1970).
80. C. K. Brown and G. Wilkinson, *J. Chem. Soc., A* p. 2753 (1970).
81. G. O. Hatta, *J. Amer. Chem. Soc.* **86**, 3903 (1964).
82. M. Orchin, *Advan. Catal.* **5**, 385 (1953).
83. I. J. Goldfarb and M. Orchin, *Advan. Catal.* **9**, 609–17 (1957).
84. V. N. Hurd and B. H. Gwynn, *World Petrol. Cong., Proc., 4th, 1955 Sect.* 4, p. 163 (1955).
85. H. L. Lemke, *Ind. Chim. (Paris)* **52**, 169 (1965).
86. A. J. Chalk and J. F. Harrod, *Advan. Organometal. Chem.* **4**, 120 (1968).
87. V. L. Hughes and I. Kirshenbaum, *Ind. Eng. Chem.* **49**, 1999 (1957).
88. I. Kirshenbaum and V. H. Hughes, *Petrol. Refiner* **37**, 209 (1958).
89. J. H. Bartlett, K. Kirshenbaum, and C. M. Menssing, *Ind. Eng. Chem.* **51**, 257 (1959).
90. W. Reppe and H. Vetter, *Justus Liebigs Ann. Chem.* **582**, 133 (1953).
91. M. Orchin, L. Kirch, and T. Goldfarb, *J. Amer. Chem. Soc.* **78**, 5450 (1956).
92. H. W. Sternberg, I. Wender, A. Friedel, and M. Orchin, *J. Amer. Chem. Soc.* **75**, 2717 (1953).
93. R. F. Heck, *J. Amer. Chem. Soc.* **87**, 4727 (1965).
94. A. I. M. Keulemanns, A. Kwantes, and T. H. Van Bavel, *Rec. Trav. Chim. Pays-Bas* **67**, 233 (1948).
95. I. Wender and H. W. Sternberg, *Advan. Catal.* **17**, 604 (1955).
96. L. Marko, *Proc. Chem. Soc., London* p. 67 (1962).
97. H. Adkins and J. L. R. Williams, *J. Org. Chem.* **17**, 980 (1952).
98. M. Morkawa, *Bull. Chem. Soc. Jap.* **37**, 379 (1964).
99. H. Uchida and A. Matsuda, *Bull. Chem. Soc. Jap.* **36**, 1351 (1963).
100. H. Uchida and A. Matsuda, *Bull. Chem. Soc. Jap.* **37**, 373 (1964).
101. C. Yokokawa, Y. Watanabe, and Y. Takegami, *Bull. Chem. Soc. Jap.* **37**, 677 (1964).
102. I. Wender, H. W. Sternberg, and M. Orchin, *Catalysis* **5**, 73 (1957).

103. R. F. Heck, *Advan. Organometal. Chem.* **4**, 243 (1966).
104. R. F. Heck, *Accounts Chem. Res.* **2**, 10 (1969).
105. J. A. Bertrand, C. L. Aldridge, S. Huseby, and H. B. Jonassen, *J. Org. Chem.* **29**, 790 (1934).
106. I. Wender, M. Orchin, and H. Storch, *J. Amer. Chem. Soc.* **72**, 4842 (1950).
107. A. Matsuda and H. Uchida, *Bull. Chem. Soc. Jap.* **38**, 710 (1965).
108. I. Wender, H. Greenfield, and M. Orchin, *J. Amer. Chem. Soc.* **73**, 2656 (1951).
109. G. Natta, P. Pino, and E. Mantica, *Chim. Ind. (Milan)* **32**, 201 (1950).
110. G. Natta, *Chim. Ind. (Milan)* **24**, 389 (1942).
111. G. Natta and E. Beati, *Chim. Ind. (Milan)* **27**, 84 (1945).
112. G. Natta and R. Ercoli, *Chim. Ind. (Milan)* **34**, 503 (1952).
113. G. Natta, R. Ercoli, S. Castellano, and F. H. Varliceri, *J. Amer. Chem. Soc.* **76**, 4049 (1954).
114. G. Natta, R. Ercoli, and S. Castellano, *Chim. Ind. (Milan)* **37**, 6 (1955).
115. A. R. Martin, *Chem. Ind. (London)* p. 1536 (1954).
116. R. Iwanaga, *Bull. Chem. Soc. Jap.* **35**, 774 (1962).
117. R. Iwanaga, *Bull. Chem. Soc. Jap.* **35**, 778 (1962).
118. R. Iwanaga, *Bull. Chem. Soc. Jap.* **35**, 247 (1962).
119. S. Metlin, I. Wender, and W. H. Sternberg, *Nature (London)* **83**, 457 (1959).
120. L. Marko, G. Bor, G. Almasy, and P. Szabo, *Bernnst.-Chem.* **44**, 184 (1963).
121. L. Kirch and M. Orchin, *J. Amer. Chem. Soc.* **81**, 3597 (1959).
122. D. S. Breslow and R. F. Heck, *Chem. Ind. (London)* p. 467 (1960).
123. R. Iwanaga, *Bull. Chem. Soc. Jap.* **35**, 865 (1962).
124. V. Macho, *Chem. Zvesti* **16**, 73 (1962); *Chem. Abstr.* **58**, 8883g (1963).
125. R. Iwanaga, *Bull. Chem. Soc. Jap.* **35**, 869 (1962).
126. Y. Takegami, C. Yoyokawa, Y. Watanabe, M. Masada, and Y. Okuda, *Bull. Chem. Soc. Jap.* **37**, 935 (1964).
127. H. Greenfield, J. H. Wotiz, and I. Wender, *J. Org. Chem.* **22**, 542 (1957).
128. P. Pino, F. Piacenti, and P. P. Neggiani, *Chem. Ind. (London)* p. 1410 (1961).
129. F. Piacenti, P. Pino, R. Lazzaroni, and M. Bianchi, *J. Chem. Soc.*, *B* p. 488 (1966).
130. Y. Takegami, C. Yoyokawa, Y. Watanabe, M. Masada, and Y. Okuda, *Bull. Chem. Soc. Jap.* **37**, 1190 (1964).
131. S. Brewis, *J. Chem. Soc. (London)* p. 5014 (1964).
132. M. Johnson, *Chem. Ind. (London)* p. 684 (1963).
133. R. W. Goetz and M. Orchin, *J. Amer. Chem. Soc.* **85**, 2782 (1963).
134. I. Wender, R. Levine, and M. Orchin, *J. Amer. Chem. Soc.* **71**, 4160 (1949).
135. R. W. Goetz and M. Orchin, *J. Org. Chem.* **27**, 3698 (1962).
136. C. L. Aldridge, E. V. Fascke, and H. B. Jonassen, *J. Phys. Chem.* **62**, 869 (1958).
137. C. L. Aldridge, U.S. Patent 2,942,034 (1960).
138. I. Wender, J. Feldman, S. Metlin, B. H. Gwynn, and M. Orchin, *J. Amer. Chem. Soc.* **77**, 5760 (1955).
139. I. Wender, S. Metlin, and M. Orchin, *J. Amer. Chem. Soc.* **73**, 5704 (1951).
140. I. Wender, H. Greenfield, and M. Orchin, *J. Amer. Chem. Soc.* **74**, 4079 (1952).
140a. H. Weber and J. Falke, *Ind. Eng. Chem.* **62**, 33 (1970).
141. M. Hill, *Petrol. Refiner*, **43**, 135 (1964).
142. A. Roche, C. Paul, and J. Saralahons, *Petrol. Refiner* **43**, 161 (1964).
143. K. Ziegler, *Chim. Ind. (Paris)* **92**, 631 (1964).
144. T. G. Selin and R. West, *J. Amer. Chem. Soc.* **84**, 1863 (1962).
145. J. L. Speier, J. A. Webster, and G. H. Barnes, *J. Amer. Chem. Soc.* **79**, 974 (1957).
146. J. C. Saam and J. L. Speier, *J. Amer. Chem. Soc.* **80**, 4104 (1958).

147. J. C. Saam and J. L. Speier, *J. Amer. Chem. Soc.* **83**, 1351 (1961).
148. K. Yamamoto and M. Kumada, *J. Organometal. Chem.* **13**, 131 (1968).
149. A. J. Chalk and J. F. Harrod, *J. Amer. Chem. Soc.* **87**, 16 (1965).
150. H. M. Bank, J. C. Saam, and J. L. Speier, *J. Org. Chem.* **29**, 792 (1964).
151. J. W. Ryan and H. Vetter, *Justus Liebigs Ann. Chem.* **582**, 133 (1953).
152. R. A. Benkeser and R. A. Hickner, *J. Amer. Chem. Soc.* **80**, 5298 (1958).
153. R. A. Benkeser, M. L. Burrous, L. E. Nelson, and J. V. Swisher, *J. Amer. Chem. Soc.* **83**, 4385 (1961).
154. A. N. Nesmayanov, R. Kh-Friedlina, E. C. Chukovskaya, R. G. Petrova, and A. B. Belyavskey, *Tetrahedron Lett.* No. 17, p. 61 (1962).
155. A. J. Chalk and J. F. Harrod, *J. Amer. Chem. Soc.* **87**, 1133 (1965).
156. A. J. Chalk and J. F. Harrod, *J. Amer. Chem. Soc.* **89**, 1640 (1967).
157. L. H. Sommer and J. E. Lyons, *J. Amer. Chem. Soc.* **90**, 4197 (1968).
158. J. S. Anderson, *J. Chem. Soc. London* p. 971 (1934).
159. J. Chatt and L. A. Duncanson, *J. Chem. Soc., London* p. 2939 (1953).
160. I. R. Joy and M. Orchin, *Z. Anorg. Allg. Chem.* **305**, 236 (1960).
161. M. J. Grogan and K. Nakamoto, *J. Amer. Chem. Soc.* **90**, 918 (1968).
162. F. Basolo, H. B. Gray, and R. G. Pearson, *J. Amer. Chem. Soc.* **82**, 4200 (1960).
163. P. M. Henry, *J. Amer. Chem. Soc.* **86**, 3246 (1964).
164. K. I. Matveev, A. M. Osipov, B. F. Odyakov, Yu. V. Suzdalnitskaya, I. F. Bukhtiyarov, and O. A. Emelyanova, *Kinet. Catal. (USSR)* **3**, 573 (1962).
165. K. I. Matveev, I. F. Bukhtiyarov, N. N. Shults, and O. A. Emelyanova, *Kinet. Catal. (USSR)* **5**, 573 (1964).
166. I. I. Moiseev, M. N. Vargaftik, and Ya. K. Sirkin, *Dokl. Akad. Nauk SSSR* **130**, 821 (1960).
167. I. I. Moiseev, M. N. Vargaftik, and Ya. K. Sirkin, *Izv. Akad. Nauk SSSR* p. 1144 (1963).
168. J. Smidt, W. Hafner, D. Jira, J. Sedlmeier, R. Siber, R. Rulliger, and H. Kojer, *Angew. Chem., Int. Ed. Engl.* **11**, 176 (1959).
169. J. Smidt and R. Seiber, *Angew. Chem.* **71**, 626 (1959).
170. J. Smidt, *Chem. Ind. (London)* p. 54 (1962).
171. M. N. Vargaftik, I. I. Moiseev, and Ya. K. Sirkin, *Dokl. Akad. Nauk SSSR* **139**, 1396 (1961).
172. M. N. Vargaftik, I. I. Moiseev, and Ya. K. Sirkin, *Izv. Akad. Nauk SSSR* p. 1147 (1963).
173. G. H. Twigg, *Chem. Ind. (London)* p. 476 (1966).
174. O. G. Levanda and I. I. Moiseev, *Kinet. Catal. (USSR)* **12**, 501 (1971).
175. P. M. Henry, *J. Amer. Chem. Soc.* **94**, 4437 (1972).
176. N. Okada, T. Noma, Y. Katsuyama, and H. Hashimoto, *Bull. Chem. Soc. Jap.* **41**, 1395 (1968).
177. J. Smidt, W. Hafner, R. Sieber, R. Sedlmeier, and A. Sabel, *Angew. Chem., Int. Ed. Engl.* **1**, 80 (1962).
178. R. Jira, J. Sedlmeier, and J. Smidt, *Justus Liebigs Ann. Chem.* **693**, 99 (1966).
179. W. Hafner, R. Jira, J. Sedlmeier, and J. Smidt, *Chem. Ber.* **95**, 1575 (1962).
180. M. S. Kharasch, R. C. Seyler, and F. R. Mayo, *J. Amer. Chem. Soc.* **60**, 88 (1938).
181. A. Aguilo, *Advan. Organometal. Chem.* **5**, 321 (1967).
182. R. Cramer, *Inorg. Chem.* **1**, 722 (1962).
183. R. R. Grinstead, *J. Org. Chem.* **26**, 238 (1961).
184. P. M. Henry, *J. Amer. Chem. Soc.* **87**, 990 (1965).
185. P. M. Henry, *J. Amer. Chem. Soc.* **87**, 4423 (1965).

186. J. E. Byrd and J. Halpern, *J. Amer. Chem. Soc.* **95**, 2586 (1973).
187. R. Creigee and C. Weiss, *Angew. Chem.* **70**, 173 (1958).
188. R. M. Flid, *Kinet. Catal. (USSR)* **2**, 58 (1961).
189. J. Halpern, B. R. James, and A. L. W. Kemp, *J. Amer. Chem. Soc.* **83**, 4097 (1961).
190. O. N. Temkin, R. M. Flid, and A. J. Malakhov, *Kinet. Catal. (USSR)* **4**, 233 (1963).
191. R. Vestin, A. Somersado, and R. Muller, *Acta Chem. Scand.* **7**, 745 (1953).
192. W. L. Budde and R. E. Dessy, *J. Amer. Chem. Soc.* **85**, 3964 (1963).
193. H. Lemaire and H. J. Lucas, *J. Amer. Chem. Soc.* **77**, 939 (1955).
194. O. N. Temkin, S. M. Brailovskii, R. M. Flid, M. P. Strukova, V. B. Belyanin, and M. G. Zaitseva, *Kinet. Catal. (USSR)* **5**, 167 (1964).
195. B. R. James and G. L. Remple, *J. Amer. Chem. Soc.* **91**, 863 (1969).
196. J. Tsuji, *Accounts Chem. Res.* **2**, 144 (1969).
197. O. N. Temkin, O. L. Kabja, G. K. Skestakov, S. M. Brailovskii, R. M. Flid, and A. P. Aseeva, *Kinet. Catal. (USSR)* **11**, 1133 (1970).
198. H. C. Clark and R. J. Puddephant, *Inorg. Chem.* **10**, 18 (1971).
199. T. Tsuji, *Accounts Chem. Res.* **6**, 8 (1973).
199a. P. M. Henry, *Accounts Chem. Res.* **6**, 16 (1973).
200. P. M. Maitlis, "The Organic Chemistry of Palladium," Vols. 1 and 2. Academic Press, New York, 1971.
200a. P. M. Maitlis, *Pure and Appl. Chem.* **33**, 489 (1973).
201. K. Ziegler, E. Holzkamp, H. Breil, and H. Martin, *Angew. Chem.* **67**, 426 and 541 (1951).
202. G. Natta, *J. Inorg. Nucl. Chem.* **8**, 589 (1958).
203. G. Natta, *Proc. Int. Conf. Coord. Chem., 3rd, 1957* p. 589 (1957).
204. G. Natta, *J. Polym. Sci.* **48**, 219 (1960).
205. K. Ziegler, *Advan. Organometal. Chem.* **6**, 1 (1968).
206. G. Natta, *Angew. Chem.* **68**, 393 (1956).
207. G. Natta, *Chem. Ind. (London)* p. 223 (1965).
208. R. Cramer, *J. Amer. Chem. Soc.* **87**, 4717 (1965).
209. R. Cramer, *J. Amer. Chem. Soc.* **89**, 1633 (1967).
210. A. D. Ketley, L. P. Fisher, A. J. Berlin, C. R. Morgan, E. H. Gorman, and T. R. Steadman, *Inorg. Chem.* **6**, 653 (1967).
211. G. Henrici-Olive and S. Olive, *J. Organometal. Chem.* **35**, 381 (1972).
212. C. A. Tolman, *J. Amer. Chem. Soc.* **92**, 6777 (1970).
213. H. Singer and G. Wilkinson, *J. Chem. Soc., A* p. 849 (1968).
214. T. Alderson, E. L. Jenner, and R. V. Lindsey, Jr., *J. Amer. Chem. Soc.* **87**, 5638 (1965).
215. J. F. Young, J. A. Osborn, F. H. Jardine, and G. Wilkinson, *Chem. Commun.* p. 131 (1965).
216. J. J. Smith, W. L. Carrick, and A. K. Ingberman, *Ann. N.Y. Acad. Sci.* **125**, 183 (1961).
217. G. Natta, A. Zambelli, G. Lanzi, I. Pasquon, E. R. Mognaschi, A. L. Segre, and P. Rentola, *Makromol. Chem.* **81**, 161 (1965).
218. L. Porri, G. Natta, and M. C. Gallazi, *Chim. Ind. (Milan)* **46**, 428 (1964).
219. J. Tsuji and S. Hosaka, *J. Polym. Sci., Part B* **3**, 703 (1965).
220. J. Tsuji and S. Hosaka, *J. Polym. Sci., Part B* **3**, 793 (1965).
221. J. P. Hermans and G. Smets, *J. Polym. Sci., Part A* **3**, 3075 (1965).
222. M. Kamachi and H. Miyama, *J. Polym. Sci., Part A* **3**, 1337 (1965).
223. G. Natta, G. D. Allasta, and G. Montroni, *J. Polym. Sci., Part B* **2**, 349 (1964).
224. G. Natta, G. D. Allasta, and L. Porri, *Makromol. Chem.* **81**, 253 (1965).

225. D. H. Bestian and K. Clauss, *Angew. Chem., Int. Ed. Engl.* **2**, 704 and 1068 (1963).
226. J. C. W. Chien, *J. Amer. Chem. Soc.* **81**, 86 (1959).
227. A. S. Matlack and D. S. Breslow, *J. Polym. Sci., Part A* 3, 2853 (1965).
228. A. Schimizu, T. Otsu, and M. Imoto, *J. Polym. Sci., Part B* 3, 449 (1965).
229. R. Bacskai, *J. Polym. Sci., Part A* 3, 2491 (1965).
230. W. P. Long and D. S. Breslow, *J. Amer. Chem. Soc.* **82**, 1953 (1960).
230a. G. Natta, G. D. Allasta, G. Mazzanti, and G. Montroni, *J. Makromul. Chem.* **69**, 163 (1963).
231. C. G. Overberger, F. S. Diachkovsky, and P. A. Jarovitzky, *J. Polym. Sci., Part A* **2**, 4113 (1964).
232. C. G. Overberger and P. A. Jarovitzky, *J. Polym. Sci., Part A* 3, 1483,(1965).
233. N. Yamazaki, K. Sasaki, and S. Kambara, *J. Polym. Sci., Part B* **2**, 487 (1964)
234. C. E. H. Bawn and R. Symcox, *J. Polym. Sci.,* **34**, 139 (1959).
235. C. E. H. Bawn and A. Ledwith, *Quart. Rev.* **16**, 341 (1962).
236. A. J. Canale, A. Hewitt, T. M. Shryne, and E. A. Youngman, *Chem. Ind. (London)* p. 1054 (1962).
237. W. L. Carrick, J. Karol, G. L. Karapinka, and J. J. Smith, *J. Amer. Chem. Soc.* **82**, 1502 (1960).
238. W. L. Carrick, W. T. Reiche, F. Penella, and J. J. Smith, *J. Amer. Chem. Soc.* **82**, 1502 (1960).
239. G. Henrici Olive and S. Olive, *Angew. Chem., Int. Ed. Engl.* **10**, 776 (1971).
240. G. Natta, G. D. Allasta, G. Mazzanti, I. Pasquon, A. Valvassori, and A. Zambelli, *J. Amer. Chem. Soc.* **83**, 3343 (1961).
241. G. W. Phillips and W. L. Carrick, *J. Amer. Chem. Soc.* **84**, 920 (1962).
242. S. Kambara, N. Yamazaki, and M. Hatano, *Amer. Chem. Soc., Div. Petrol. Chem., Prepr.* A23 (1964).
243. P. Cossee, *J. Catal.* 3, 80 (1964).
244. W. L. Carrick, *J. Amer. Chem. Soc.* **80**, 6455 (1958).
245. F. J. Karol and W. L. Carrick, *J. Amer. Chem. Soc.* **83**, 2654 (1961).
246. P. Cossee, *Tetrahedron Lett.* No. 17, pp. 12 and 17 (1960).
247. W. L. Carrick, R. W. Kluiber, E. F. Bonner, L. H. Wartman, F. M. Rugg, and J. J. Smith, *J. Amer. Chem. Soc.,* **82**, 3882 (1960).
248. P. Cossee, *in* "Advances in the Chemistry of Coordination Compounds" (S. Kirschner, ed.), p. 24. Macmillan, New York, 1961.
249. J. Chatt and B. L. Shaw, *J. Chem. Soc., London* p. 705 (1959).
250. P. R. H. Alderman, P. G. Owston, and J. H. Rowe, *Acta Crystallogr.* **13**, 149 (1960).
251. S. Ikeda and R. Tsuchiya, *J. Polym. Sci., Part B* 3, 39 (1965).
252. D. S. Breslow and N. R. Newburg, *J. Amer. Chem. Soc.* **81**, 81 (1959).
253. M. H. Lehr and P. H. Moyer, *J. Polym. Sci., Part A* 3, 231 (1965).
254. P. H. Moyer and M. H. Lehr, *J. Polym. Sci., Part A* 3, 217 (1965).
255. M. Modena, R. B. Bates, and C. S. Marvel, *J. Polym. Sci., Part A* 3, 949 (1965).
256. G. Natta, L. Porri, A. Carbonaro, and G. Stoppa, *Makromol. Chem.* 77, 114 (1964).
257. G. Natta, L. Porri, and A. Carbonaro, *Makromol. Chem.* **67**, 126 (1964).
258. H. Noguchi and S. Kambara, *J. Polym. Sci., Part B* **2**, 593 (1964).
259. K. Matsuzaki and T. Yasukawa, *J. Polym. Sci., Part B* 3, 393 (1965).
260. K. Matsuzaki and T. Yasukawa, *J. Polym. Sci., Part B* 3, 907 (1965).
261. A. Takahashi and S. Kambara, *Polym. Sci., Part B* 3, 279 (1965).
262. J. G. Balas, H. E. Delamare, and D. O. Schissler, *J. Polym. Sci., Part A* 3, 2243 (1965).

263. A. I. Deaconescu and S. S. Medvedev, *J. Polym. Sci., Part A* **3**, 31 (1965).
264. W. Reppe, O. Schlichting, K. Klager, and T. Toepel, *Justus Liebigs Ann. Chem.* **560**, 1 (1948).
265. R. E. Rinehart, H. P. Smith, H. S. Witt, and H. Romeyn, Jr., *J. Amer. Chem. Soc.* **83**, 4867 (1961).
266. R. E. Rinehard, H. P. Smith, H. S. Witt, and H. Romeyn, Jr., *J. Amer. Chem. Soc.* **84**, 4145 (1962).
267. P. Teyessie and R. Dauby, *J. Polym. Sci., Part B* **2**, 413 (1964).
268. M. Tsutsui and J. Ariyoshi, *J. Polym. Sci., Part A* **3**, 1729 (1965).
269. G. Wilkie, *Amer. Chem. Soc., Div. Petrol. Chem., Prepr.*, A67 (1964).
270. G. Natta, L. Porri, and S. Valenti, *Makromol. Chem.* **67**, 225 (1963).
271. G. Natta, L. Porri, A. Carbonaro, and A. Greco, *Makromol. Chem.* **77**, 207 (1964).
272. C. E. H. Bawn, D. G. T. Cooper, and A. M. North, *Polymer* **7**, 113 (1966).
273. E. Ochiai, H. Hirani, and S. Makashima, *J. Polym. Sci., Part B* **4**, 1003 (1966).
274. I. V. Nicolescu and E. Angelescu, *J. Polym. Sci., Part A* **3**, 1227 (1965).
275. L. S. Meriwether, E. C. Colthup, G. M. Kennerly, and R. N. Reusch, *J. Org. Chem.* **26**, 5155 (1961).
276. L. S. Meriwether, M. F. Leto, E. C. Colthup, and G. W. Kennerly, *J. Org. Chem.* **27**, 3930 (1962).
277. A. Yamamoto, K. Morifuji, S. Ikeda, T. Saito, Y. Uchida, and A. Misono, *J. Amer. Chem. Soc.* **87**, 4652 (1965).
278. R. J. Kern, *Chem. Commun.* p. 706 (1968).
279. C. K. Brown, D. Georgion, and G. Wilkinson, *J. Chem. Soc., A* p. 3120 (1971).
280. R. F. Heck, *Advan. Chem. Ser.* **49**, 181 (1965).
282. S. Tanaka, K. Makuchi, and N. Shimazaki, *J. Org. Chem.* **29**, 1626 (1964).
283. S. Otsuka, T. Taketomi, and T. Kikuchi, *J. Amer. Chem. Soc.* **85**, 3709 (1963).
284. D. Wittenberg, *Angew. Chem., Int. Ed. Engl.* **3**, 153 (1964).
285. H. Muller, D. Wilinby, H. Seilut, and E. Scharf, *Angew. Chem., Int. Ed. Engl.* **4**, 327 (1965).
286. M. H. Hidai, Y. Uchida, and A. Misono, *Bull. Chem. Soc. Jap.* **38**, 1243 (1965).
287. T. Saito, Y. Uchida, and A. Misono, *Bull. Chem. Soc. Jap.* **37**, 105 (1964).
288. T. Saito, Y. Uchida, and A. Misono, *Bull. Chem. Soc. Jap.* **38**, 1397 (1965).
289. J. Feldsman, O. Frampton, B. Saffer, and H. Thomas, *Amer. Chem. Soc., Div. Petrol. Chem., Prepr.* A33 (1964).
290. M. Tsutsui and J. Ariyoshi, *Trans. N.Y. Acad. Sci.* [2] **26**, 431 (1964).
291. S. Takahashi, T. Yamazaki, and N. Hagihara, *Bull. Chem. Soc. Jap.* **41**, 254 (1968).
292. A. Misono, Y. Uchida, M. Hidai, and I. Inomata, *Chem. Commun.* p. 704 (1968).
293. A. Misono, Y. Uchida, M. Hidai, H. Shinohara, and Y. Watanabe, *Bull. Chem. Soc. Jap.* **41**, 396 (1968).
294. A. Carbonaro, A. Greco, and G. D. Allasta, *Tetrahedron Lett.* No. 22, p. 2037 (1967).
295. A. Carbonaro, A. L. Segre, A. Greco, C. Tosi, and G. D. Allasta, *J. Amer. Chem. Soc.* **90**, 4453 (1968).
296. R. Cramer, *Accounts Chem. Res.* **1**, 186 (1968).
297. S. Takahashi, T. Shihano, and N. Hagihara, *Bull. Chem. Soc. Jap.* **41**, 459 (1968).
298. M. Tsutsui, *Trans. N.Y. Acad. Sci.* [2] **26**, 423 (1964).
299. M. Tsutsui, *Ann. N.Y. Acad. Sci.* **125**, 147 (1965).
300. E. Ochiai, *Coord. Chem. Rev.* **3**, 49 (1968).

301. B. A. Dolgoplosk, I. A. Oreshkin, E. I. Tinyakova, and V. A. Yakoolev, *Bull. Acad. Sci. USSR* 9, 2059 (1967).

302. G. Wilkie, B. Bogdanovic, P. Hardt, P. Heimbach, W. Keim, M. Kroner, D. Oberkirch, K. Tanaka, E. Steinrucke, D. Walter, and H. Zimmerman, *Angew. Chem., Int. Ed. Engl.* 5, 151 (1966).

303. G. Wilkie, *Angew. Chem., Int. Ed. Engl.* 2, 105 (1963).

304. G. N. Schrauzer, *Angew. Chem., Int. Ed. Engl.* 3, 185 (1964).

305. G. N. Schrauzer, *Advan. Organometal. Chem.* 2, 2 (1964).

306. M. Tsutsui and H. Zeiss, *J. Amer. Chem. Soc.* 82, 6225 (1960).

307. G. N. Schrauzer, *J. Amer. Chem. Soc.* 81, 5310 (1959).

308. G. N. Schrauzer, *Chem. Ber.* 94, 1403 (1961).

309. J. P. Collman, J. W. Kang, W. F. Luile, and F. Sullivan, *Inorg. Chem.* 7, 1298 (1968).

310. J. P. Collman, *Accounts Chem. Res.* 1, 136 (1968).

311. H. Reinheimer, H. Dietl, J. Moffat, D. Wolff, and P. M. Matilis, *J. Amer. Chem. Soc.* 90, 5321 (1968).

312. H. Reinheimer, J. Moffat, and P. M. Maitlis, *J. Amer. Chem. Soc.* 92, 2285 (1970).

Appendix I

Formulas, Bonding, and Formal Charges

The conventions used in representing structural formulas of metal–organic complexes are designed to show the distribution of bonding electrons in a reasonably consistent fashion, as well as to make possible completely unambiguous "electron-bookkeeping" of bonding electrons. An ideal method of representation is impossible for the wide variety of metals and ligands encountered in this book, since seeming inconsistencies will develop for metals or donors at the extremes of the reactivity scale. Thus a consistent way of representing a donor–acceptor (coordinate) bond for electropositive metal ions (e.g., Al^{3+}), with the electrons localized on the electronegative atom, would be misleading for the complexes of highly electronegative metal ions (e.g., Hg^{2+}, Pt^{2+}) for which linkages with the same donor would be highly covalent, with sometimes very little polarity. Thus a consistent way of writing coordinate bonds will lead to inconsistencies in representation of formal charge distribution, which the reader must continually keep in mind, and make allowances for, if this (or any similar) book is to be interpreted correctly.

The objective of accounting for all bonding electrons by the way that bonds and formal charges are represented also provides an automatic method of obtaining the charges of all ionic species. Thus the total charge on a complex ion is equal to the sum of the formal charges indicated at various positions in the formulas. In many instances the total charge is not indicated, but may be deduced by summation of the indicated formal charges. For covalent (homopolar) bond representations, the bonding electrons are (arbitrarily) assumed to be shared equally between the two atoms involved. Similarly, for coordinate (donor–acceptor) linkages, the formal charges are

177

determined by assuming that the bonding electrons are localized on the electronegative donor atom.

The conventions employed for homopolar and coordinate bonding in structural and graphic formulas are shown in the following tabulations. While the isoelectronic NO^+ and CO groups are shown with solid bonds to metal, the negative NO^- group is considered a donor ligand and is represented with dashed lines for coordinated bonding to metal ions.

COVALENT BONDS (SOLID LINE)

Type	Example	
Carbon–carbon	H_3C—CH_3	Solid line
Metal hydride	M—H	Solid line
Metal alkyl (σ-bond)	M—CH_3	Solid line
Metal carbonyl	M—C≡O	Solid line
Metal nitrosyl	M—N≡O^+	Solid line

DONOR-ACCEPTOR BONDS

Type	Example	
Metal complex (with electronegative atom of ligand)	M^{n+}----Cl^- M^{n+}----NH_3	Broken line
Metal alkene or alkyne complex (π bond)	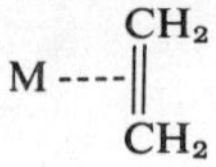 M----‖ (with CH_2 above and CH_2 below)	Broken line
Metal–carbon bond formed by an electronegative ion or group	^-NC----Ag^+----CN^-	Broken line
Transition state (partial bonds)	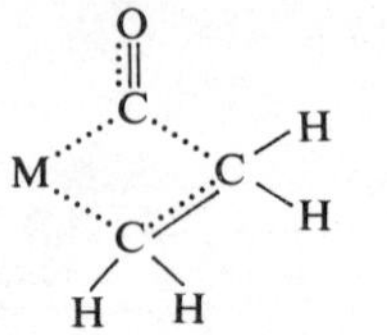	Heavy dotted line
Outline of figures (no bonds)	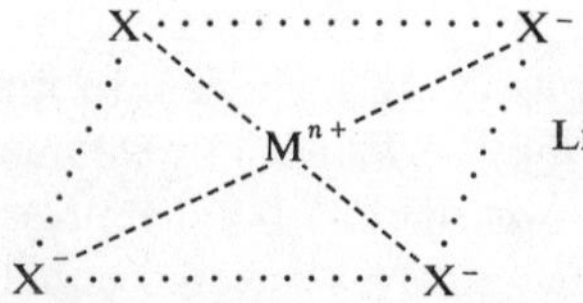	Light dotted line

More complex types of compounds are frequently represented by a simple designation, such as a single solid line to represent a π-allyl complex, as represented below.

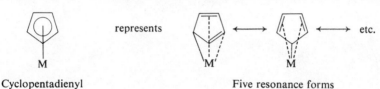

π-Allyl complex   represents   Resonance forms

Similarly a sandwich-type complex or a π-complex with an organic aromatic ring is usually represented by a single bond as indicated below.

represents   etc.

Cyclopentadienyl
metal complex   Five resonance forms

Appendix II

Glossary of Terms and Abbreviations

Abbreviation	Full name or formula
Ac	Acetyl
acac	2,4-Pentanedione
AcO	Acetate
ADP	Adenosine diphosphate
$AsPh_3$	Triphenylarsine
ATP	Adenosine triphosphate
B	Base
BPh_4	Tetraphenylboron anion
$BiPh_3$	Triphenylbismuth
CDTA	Cyclohexanediaminetetraacetic acid
Cobaloxime-II	Bis(dimethylglyoximato)cobalt(II)
Cobalamine	Vitamin B_{12}
cp	π-Cyclopentadienyl
das	*o*-Phenylenebis(dimethylarsine)
depe	1,2-Bis(diethylphosphino)ethane
dias	1,2-Bis(diphenylarsino)ethane
diars	1,2-Bis(methylphenylarsino)ethane
dien	Diethylenetriamine
diphos	1,2-Bis(diphenylphosphino)ethane
diphos-2	*cis*-1,2-Bis(triphenylphosphino)ethylene
dipy	2,2-Dipyridyl
dmpe	1,2-Bis(dimethylphosphino)ethane
DMF	Dimethylformamide
DMSO	Dimethyl sulfoxide
DPN	Diphosphopyridine nucleotide
DPNH	Reduced DPN
DTPA	Diethylenetriaminepentaacetic acid
EDTA	Ethylenediaminetetraacetic acid

Glossary (*continued*)

Abbreviation	Full name or formula
en	Ethylenediamine
epr	Electron paramagnetic resonance
fac	Facial
Fenton's reagent	Hydrogen peroxide and an iron(II) salt
glygly(GG)	Glycylglycine
HEDTA	N-Hydroxyethylethylenediamine-N',N'-triacetic acid
HIMDA	N-Hydroxyethyliminodiacetic acid
HPIP	High potential iron protein
IMDA	Iminodiacetic acid
i-pr, pr^i	Isopropyl
L	Ligand
mer	Meridional
NAD	Pyridineadenine dinucleotide
NADH	Reduced pyridineadenine dinucleotide
NMR	Nuclear magnetic resonance
n-pr, pr^n	Normal propyl
ox	Oxidized
pm	Primary
PMR	Proton magnetic resonance
PPh$_3$	Triphenylphosphine
PTS	Phthalocyaninetetrasulfonic acid
Py	Pyridine
rac	Racemic
red	Reduced
S	Solvent
salen	N,N'-Ethylenebis(salicylideneiminato)
SbPh$_3$	Triphenylantimony
sec	Secondary
synthesis gas	Equimolar carbon monoxide and hydrogen
t-Bu or Bu^t	Tertiary butyl
tert	Tertiary
THF	Tetrahydrofuran
tolan	Diphenylacetylene
TPN	Triphosphopyridine nucleotide
TPNH$_2$	Reduced TPN
trien	Triethylenetetraamine
Vaska's complex	Chlorocarbonylbis(triphenylphosphine)iridium(I)
Wilkinson's complex	Chlorotris(triphenylphosphine)rhodium(I)
Zeise's salt	Potassium trichloro(ethylene)platinate(II) monohydrate

Author Index

Numbers in parentheses are reference numbers and indicate that an author's work is referred to although his name is not cited in the text. Numbers in italics show the page on which the complete reference is listed.

A

Adkins, H., 42(97), *170*
Aguilo, A., 83, *172*
Alderman, P. R. H., 116(250), *174*
Alderson, T., 100(214), 139, 139(214), 141 (214), 143(214), 146(214), *173*
Aldridge, C. L., 45, 50, 61, 61(137), *171*
Allasta, G. D., 101(223, 224), 102(240), 103(223, 224), 104(230a), 107, 107(224), 108(224), 141(294, 295), 144(294, 295), 148(294, 295), *173, 174, 175*
Almasy, G., 49, 51, *171*
Anderson, J. S., 77, 92(158), *172*
Angelescu, E., 129(274), 130(274), *175*
Ariyoshi, J., 119(268), 123(268), 128(268), 141(290), 143(290), 151(290), *175*
Arnet, J. E., 17, 17(38), *169*
Aseeva, A. P., 96(197), *173*

B

Bacskai, R., 102(229), 104(229), *174*
Bailar, J. C., 31(61), 34(61), *170*
Balas, J. G., 119(262), 123(262), 126(262), 127(262), *174*

Ballhausen, C. J., 3(9), *168*
Bank, H. M., 72, *172*
Barnes, G. H., 66, *171*
Bartlett, J. H., 40(89), *170*
Basolo, F., 78(162), *172*
Bates, R. B., 119(255), 120(255), *174*
Bawn, C. E. H., 102(234), 105, 105(235), 126, 127(272), *174, 175*
Beati, E., 48(111)
Beck, W., 13(28), *169*
Belyanin, V. B., 94(194), *173*
Belyavskey, A. B., 73, 73(154), *172*
Benkeser, R. A., 72, 72(153), *172*
Berlin, A. J., 100(210), *173*
Bertrand, J. A., 45, 50, *171*
Bestian, D. H., 104(225), 113(225), 114 (225), 136(225), 148(225), *174*
Bianchi, M., 53, *171*
Bodanovic, P., 159(302), *176*
Bond, G., 25, 31, 32, 32(49, 50), 33(49), 36, *169*
Bonner, E. F., 114(247), *174*
Bor, G., 49, 51, *171*
Brailouskii, S. M., 94(194), 96(197), *173*
Braun, G., 13(28), *169*

183

Subject Index

191